THE LOGIC OF DISCOVERY

SYNTHESE LIBRARY

STUDIES IN EPISTEMOLOGY,

LOGIC, METHODOLOGY, AND PHILOSOPHY OF SCIENCE

VOLUME 231

SCOTT A. KLEINER
Department of Philosophy, University of Georgia, Athens, U.S.A.

THE LOGIC OF DISCOVERY

A Theory of the Rationality of Scientific Research

Springer-Science+Business Media, B.V.

Library of Congress Cataloging-in-Publication Data

Kleiner, Scott A.
The logic of discovery : a theory of the rationality of scientific research / by Scott A. Kleiner.
p. cm. -- (Synthese library ; v. 231)
Includes index.

1. Research--Methodology. 2. Research--Philosophy. 3. Logic.
I. Title. II. Series.
Q180.55.M4K58 1993
507.2--dc20 93-22798

ISBN 978-90-481-4297-2 ISBN 978-94-015-8216-2 (eBook)
DOI 10.1007/978-94-015-8216-2

Printed on acid-free paper

TO:

L.W. Scott Alter

Ruth Shippen Alter

Margaret Shippen Kleiner

AND TO:

Heather S. Kleiner

Catherine B. Kleiner

Margaret K. Cederberg

whose loving persistence at various stages of my life was necessary for the appearance of this work

CONTENTS

PREFACE

This book addresses several related philosophical problems, perhaps the most important of which are: What is the logic of scientific discovery? Once an answer is proposed for this question, further commonly asked questions may be considered: Can philosophers say anything illuminating about scientific discovery? Can we expect only illumination, or can inquiry into various aspects of discovery actually aid the discovery process? The first of these questions divides into further subquestions: What is scientific discovery? What 'logic' is appropriately attributed to scientific discovery? Furthermore plausible answers to some of these questions generate further questions: If scientific discovery is novel scientific knowledge, what constitutes such novelty? For example is it the occurrence of new theories or practices in science that are incommensurably different from their predecessors, or does scientific novelty depend in an essential way upon what has occurred in the past? As to 'logic', are there, or can philosophers, scientists or others create routines for generating scientific knowledge? Does deduction play a role in generating scientific novelty? What role, if any, does induction play, and what is meant by "induction"? Are there other principles, rules or norms that play an essential role in guiding investigators in generating scientific novelty, and if so, what are they? How do they differ from deductive principles, inductive procedures, and algorithmic operations? By what criteria and by what principles are they appraised? It is hoped that answers to the above questions will shed new light on an old question: What is scientific rationality and what are the limits within which scientific rationality can appropriately operate? For example, are there rational ways to generate as well as rational ways to appraise items that are put forth as candidates for scientific knowledge? Is the rationale of appraisal distinct from that of generation, or are the two essentially and importantly related?

My interest in this topic came as a byproduct of a long standing interest in the logic of scientific questions. I read Sylvain Bromberger's 'A Theory about the Theory of Theory and about the Theory of Theories' as a graduate student in the early 1960's, about the same time I was exposed to T.S. Kuhn's *Structure of Scientific Revolutions.* Kuhn's claim that paradigms include an endemic set of questions or problems suggested to me that the logic of questions might help to further illuminate the normal scientific traditions and the scientific revolutions described by Kuhn. My Ph.D. dissertation, *The Concept of Force and the Explanatory Efficacy of Newtonian Mechanics* was an attempt to describe the explanatory enterprise of Newtonian mechanics as a program of research that could be formulated in terms of a cluster of questions. In another paper (1970) I attempted to show the logical linkage between the world view constituting a paradigm and the questions promulgated therein. This effort involved me in problems of meaning and incommensurability, to which I turned for the next ten years. Then in 1980 an invitation from Jaakko Hintikka revived my interest in questions. An encounter with Larry Laudan's *Progress and Its Problems* convinced me that scientific problems are important and that there is a methodology of problem choice that goes beyond the falsification that can be explained by erotetic logic. Also reading Gruber and Barrett's *Darwin on Man* instilled an interest in the discovery process and raised questions as to how and to what extent it could be considered deliberate activity and thus subject to some kind of logic. Thomas Nickles' conference proceedings, *Scientific Discovery, Logic and Rationality* nourished this interest. At about the same time Bill Wimsatt encouraged me to look at heuristics, which led me to Herbert Simon's *Models of Discovery* and the various efforts to model the discovery process in artificial intelligence.

I have attempted to make this work as self-contained as is possible within a finite volume so that the more technical items of logic, computational inquiry and scientific theory should be accessible without prior reading in these areas. These materials should be accessible to any interested scientist as well as to readers with some undergraduate philosophy and interests in the methods of scientific inquiry. They are appropriate as part of an advanced undergraduate or a graduate course in Philosophy of Science, but should not be considered as sufficient for providing an overview of the discipline. Though each chapter can probably be understood independently of the others, I believe the book should be studied in its entirety to appreciate the various aspects of the discovery process in science. Only the final chapter, Chapter 6, may be considered as independent of the rest of the work. Each chapter is followed by a short summary and conclusion which should give a fairly comprehensive overview of the topics and arguments presented and discussed therein.

Chapter 1 introduces the problem of discovery and discusses various motives for avoiding it in both Logical Empiricist and Historicist traditions. In Chapter 2 the ontological, conceptual, metascientific and empirical components of scientific belief systems are distinguished as independent variables in scientific development and their role in the formulation and choice of scientific questions and the pursuit of answers is discussed. Chapter 3 contains an exposition of the main components of Aqvist's and Hintikka's deontic and epistemic formulation of erotetic logic and a discussion of how this logic can represent conceptual change, the falsification and the correction of questions. Here I also make a number of initial points regarding the limitations of erotetic logic as a means of modeling scientific inquiry. Chapter 4 is devoted to the problem of question preference, both as can be dealt with within the confines of erotetic logic and by heuristic and other considerations external to this logic, including applications of probability theory and Bayes' theorem and Laudan's criteria for weighting problems. Chapter 5 is concerned with heuristics and the pursuit of problems, and discusses problems as construed by researchers in artificial intelligence, as well as the presuppositions, strategies and capabilities of Simon's BACON programs. These programs are contrasted with another approach to problem solving based upon substantive knowledge and a network of relevance relations in the PROCESS OF INDUCTION programs discussed by Paul Thagard and others. The latter programs provide examples of several further heuristics for which methods of appraisal are discussed as components of a logic of discovery. Finally in Chapter 6 various traditionally advocated methods of discovery, the various forms of induction, hypothetico-deductive methods, abduction, and the method of analysis are discussed and construed as some among many scientific heuristics.

Much of the text was written while I was a fellow at the Center for the Philosophy of Science at the University of Pittsburgh in the fall of 1985 and the winter of 1987. I am indebted to many of the fellows in residence at these times for helpful discussions on various aspects of the topic, particularly to Nancy Nersessian, Ulrich Majer, Lothar Schafer, Diderik Batens, and Timo Airaksinen. I am also indebted to Nicholas Rescher for useful suggestions and Peter Machamer for criticisms of my initial use of epistemic erotetic logic. A referee for this series offered several criticisms and helpful suggestions. Donald Nute has been most generous in offering the typesetting facilities of the AI lab at the University of Georgia, and Jessie Liu must be thanked for so willingly and competently doing the typesetting routines and much of the busywork in generating the index.

CHAPTER ONE

INTRODUCTION

1.1. RECENT PHILOSOPHY OF SCIENCE AND DISCOVERY

Most twentieth century philosophy has avoided any attempt to consider creativity, invention, novelty or discovery in science. Some philosophers, following Einstein, hold that novelties in science are 'free creations of the mind' for which there should be no definitional, constructive or other restrictive logical ties to what is actually known at the observational level (Popper, 1958, Einstein, 1953 , Hempel, 1966; but also Feyerabend, 1974, 1987).

Others hold that new ideas emerge through processes of illumination, which may include a portion of unconscious processing of ideas, for which there may be a descriptive psychology but no normative logic. This psychologistic view, *illuminism*, implies that the production of conceptual novelty is beyond the deliberate control of human inquirers, that normative guidelines for inquiry directed to conceptual novelty are thus inappropriate presumably because we cannot act to direct, guide or constitute mental processing, and the best the inquirer can do is to wait and hope to be blessed with novel ideas. So described, the discovery process, or the creative processes of scientific research, cannot be guided by preferred objectives or values, rules or strategies, and thus would be a-rational. Polanyi (1958) agrees that 'happy moments' of discovery cannot be engineered at will but he goes on to say that the stage is set for such occurrences by deliberate actions such as the dedicated pursuit of a problem and by suitably timed rest periods for the 'incubation' of novel ideas. Thus one may act to prepare herself for the appreciation of the significance of such illumination if not to make their occurrence more probable. Simon (1977) offers hypotheses about what happens in the unconscious processing that may produce 'illuminations' after periods of repose, but he also holds that this processing can be carried out consciously and deliberately, and thus a-rational illuminism cannot be attributed to him. Once the algorithms underlying this subconscious processing are known they may be applied deliberately under appropriate circumstances.

Another more recent view is that significant scientific novelty, viz. scientific revolutions, are holistic in their scope and thus include changes in concept and method as well as substantive belief. Since scientific revolution includes changes in the aims and problems of a discipline, in the means or procedures for solving problems, as well as in the world its inquirers believe themselves to be studying, revolutionary changes cannot be guided by normative epistemic standards or by substantive beliefs characteristic of a prior tradition. Rather, revolutionary research is more like random exploration (Kuhn, 1970) and may even run contrary to established methodological norms (Feyerabend, 1974). However, others have questioned the holism of scientific revolutions (Laudan, 1984). If revolutions are not holistic the possibility remains open that revolutionary scientific research can be rationally guided by some paradigmatic belief, whether substantive, metascientific or valuational, that remains unchallenged and unchanged in the revolutionary process.

Finally, there is the belief that scientific novelty is serendipitous, that is, it cannot be intended, planned or deliberately sought. Rather novelty always consists of the unexpected: The pursuit of established problems can generate conceptions that eliminate these problems, direct attention to new problems or solve problems whose solution was neither sought nor anticipated. Genuinely novel ideas evolve or are produced independently of the problems they solve, or independently of the pursuit of those problems (Kantorovitch and Ne'eman, 1989; See also Kuhn, 1970). A defense of the rationality of discovery need not deny the occurrence of fortuitous and serendipitous discoveries. One need only acknowledge that though one's objectives and guidelines for inquiry may be mistaken or inappropriate, planned epistemic research based upon ones best considered judgments is better than passive agnosticism. Furthermore epistemic novelties or surprise encounters are generally appreciated only by 'prepared' minds, that is, individuals with relevant substantive beliefs and research agendas. Often the significance of the unexpected is in the revisions entailed among the worldviews, paradigmatic achievements, values and methods deliberately employed by an investigator.

The arguments against a logic of discovery are in part addressed against Baconian 'inductivists' and in part against Cartesian rationalists, to whom the hope is attributed that epistemological analysis of science would produce universally applicable routines for creating good science that could be undertaken by anyone, regardless of her intellectual gifts. Popper (1958), Hempel (1966) and other defenders of the 'hypothetico-deductive method' argued at length that no such routines could have generated the powerful theoretical concepts of physics and other branches of natural science. Arguments against normative procedures for discovery are also products of logical positivist and logical empiricist philosophers who, in extending metamathematical thinking to the philosophy of empirical science, are inclined to believe that there are logical procedures for checking or justifying scientific beliefs but not for generating them. Also they claim that these logical procedures are only properly applicable to completed, if not formalized, scientific theories (Carnap, 1953). Furthermore it is argued that the source of an idea or its history has nothing to do with its credibility or its content. At best the considerations that produced a discovery can give it only provisional initial support and the best grounds for accepting it come subsequently with unexpected success or with success in dealing with problems that were initially deemed unfavorable (Herschel, 1831, Whewell, 1840, Popper, 1958, 1963, Hanson, 1961, Salmon, 1966, Lakatos, 1970, Laudan, 1980). Others in the positivist and logical empiricist traditions assume that possible and appropriate evidence for a given proposition or conception can be obtained exclusively by logical analysis of constituent concepts, whereas origins and developmental history are entirely contingent. Depending upon historical and social circumstances there can be many possible routes to an item of scientific knowledge but only one ideally appropriate justificatory procedure. Paths of discovery and evidence according to this view are wholly disjoint.

Furthermore, it is sometimes argued that genuine conceptual novelty cannot be anticipated and yet remain conceptual novelty. Anticipating an idea is having the idea. Hence all novel ideas must come as surprises. They cannot be deliberately sought, though they might be encountered by pursuing the idea that they ultimately replace (Kuhn, 1970).

Finally, one philosopher, namely Feyerabend (1974), has argued that established research traditions cannot be reasonably opposed. Such opposition must always run counter to established methodology or scientific rationality, which is inherently conservative. The production of novelty is only opposed by methodological norms, and these norms must be violated in the propagation of novel ideas, which is done best by the pursuit of fancy and the production of community consensus by a-logical propaganda.

Philosophers in the 'historicist' tradition, particularly Hanson (1961), Toulmin (1961) and Kuhn (1970) have recommended, against the positivistic tradition, that we should stop considering science in its ideally finished condition. This condition consists of axiomatic formulation of theories using well-established concepts perhaps purified by logical reconstruction, and known links to known classes of phenomena. In the positivistic tradition axiomatization is the point at which philosophical inquiries begin, e.g. inquiries into the epistemically optimal nature of law, theory, explanation, reduction, or confirmation. Rather, historicists claim, a great deal more can be learned about the problems faced in the actual practice of science if scientific concepts are thought of as dynamic or in a state of flux and if science is thought of as an activity or an enterprise in which various goals are pursued by appropriately chosen means. These recommendations have encouraged philosophers to pay close heed to the history of science and particularly to differences in substantive belief, concept and method appearing in various stages in the development of a discipline. They have raised new methodological problems, particularly those of the rationality of scientific revolutions, which are episodes in which prevailing methodologies, worldviews, concepts, and observational procedures have more or less radically changed.

However, much of this work is retrospective reconstruction in which past obsolete theories, as well as current theories, are considered in their final or near-final published form, with little attention to the activities which produced them. With some exceptions (most notably Hanson), the more influential writers of this tradition (particularly Kuhn, Lakatos, Feyerabend and Laudan) have made significant contributions only to the understanding of certain kinds of scientific activity, particularly what Kuhn calls normal science or Lakatos' established scientific 'research programmes'. Generally they have denied that scientific research ever deliberately aims at or systematically pursues fundamental novelty. They have also denied that there are any circumstances in which 'normal programs' have reached a point of decisive 'degeneration' wherein the search for novelty can be rationally justified. As have many others, they have either explicitly claimed, or allowed their audiences to assume by default, that significantly novel ideas either occur by accident or by non-logical illumination. When they do occur, programs for their development, if needed at all, must proceed against the weight of epistemologically sanctioned rationality. As in the case of the logical empiricist tradition, the most that certain dissenters could say for discovery is that it is hindered by overly restrictive logical or methodological demands (compare Feyerabend, 1974, with Hempel, 1966, or Popper, 1958).

1.2. SOME MISCONCEPTIONS REGARDING DISCOVERY

The aim of this essay in part is to show that discovery, or the production of scientific novelty, when properly understood, is at least as important an area of philosophic concern as is the more traditional epistemological endeavor of determining criteria for scientific intelligibility and credibility. In this section a few remarks will be made about what a possible logic of discovery is not. These observations will not solve the problem of producing an adequate conception of rational scientific inquiry, though they are first steps in that direction.

In his 'Rules for the Direction of the Mind' Descartes described what might be taken to be a logic of discovery:

> ...by a method I mean certain and simple rules, such that, if a man observe them accurately, he shall never assume what is false as true, and will never spend his mental efforts to no purpose, but will always gradually increase his knowledge and so arrive at a true understanding of all that does not surpass his powers. (Descartes, 1927, p. 49)

The Cartesian ideal for scientific method is a routine that even the untalented can use with success by mechanical application, and it is sufficient for generating all we are capable of knowing. This ideal is in general unrealizable for many important aspects of inquiry in natural science. The contemporary equivalent of the Cartesian search is the design of algorithms by which computers can produce outputs of scientific importance. It is well known that such 'effective' procedures, or equivalently the recursive application of deterministic computational rules, are sufficient for certain computational routines, such as summing, multiplying, taking roots, computing logarithms, but fall short of sufficiency for many other problems of scientific importance.

Some developers of machine intelligence (e.g. Simon, 1977, Nilsson, 1972, Langley, Simon, *et. al.*, 1987, Thagard, 1986, Holland, *et. al.,* 1988) have construed scientific discovery as the solution of problems. By their conception problems present goals to be achieved, and the effort to achieve such a goal involves a search through a space defined by the problem. A problem space is a symbolic representation in the brain, a kind of conceptual structure that includes initial, intermediate and final 'problem situations', which are cognitive states. The problem space may be restricted by background beliefs as well as beliefs characterizing the final or goal state. The search begins in a portion of that space called an initial state and generates algorithmically from that state further locations in the hope of arriving at a state which solves the problem, that is a belief that satisfies the description of the goal state. However, from each location in the problem space a large number of options are available for moving to subsequent locations, and, accordingly, such search programs have the power of generating an exponential explosion of possible paths of exploration. Under these conditions it is not practicable for even high speed computers to reach a solution. To head off such explosions, designers of search programs have sought guidelines or *heuristics* which suggest plausible actions to undertake and plausible ones to avoid in specific circumstances. By guiding the search at each step heuristics can restrict the exponential increase of possible steps as one follows the path of a search for a solution to a problem. Heuristics thus facilitate a search by making practically possible or probable the production of a solution, but they do not guarantee the production of the outcome sought. Also they can be expected to be successful only in certain circumstances (Wimsatt, 1980, Nickles, 1990). A universal heuristic of the Cartesian sort is not what is being sought in the pursuit of artificial intelligence, and, analogously, should not be sought for science. There are many such heuristics already known to researchers in artificial intelligence, and there are surely many more to be discovered. However machine intelligence has as yet been successful only with certain rudimentary problems in relatively well-defined conditions of background knowledge, specificity of problem or objective. Its applications include the diagnosis of disease, the determination of molecular structure, determining functional relations from tabulated numerical data, recognizing visual patterns, and drawing inferences from analogy. Those who offer these programs as exemplars of a 'logic of discovery' have not made clear why the products of these programs should be called 'discoveries' or why the heuristics they employ should be considered a 'logic'. These procedures will be discussed in more detail in Chapter 3 below. Also, scientists are often faced with problems of choosing epistemic objectives and paths of research where the problem itself and relevant background knowledge is ill-defined, various or lacking in consensus, or of questionable importance (Simon, 1977). Though the hardware and software of machines is being made increasingly more 'talented' and thus able to perform more sophisticated tasks, there is hardly apparent in these developments a reduction of the many tasks of science to heuristics utilizable by machines, such as the task of justifying a problem or finding new ways of providing evidence for a belief. Nor is it clear such reductions should be sought, for they depend on the debatable premise that all human minds, including scientific minds of the present and the past, operate on programs which in principle can be written for machines whose design and construction is in principle possible. A weaker and perhaps more plausible position is that computing machines promise to free the inquirer of tasks whose solutions were initially not routine, but became routine though perhaps time consuming for human beings. Then human time and talents can be spent on formulating, choosing, and searching for methods for solving problems beyond those whose solution has been mechanized. In some cases these routines can be for the production of novelty on both the scientific level, e.g. the discovery of constants and functional relations from empirical data, and the metascientific level, e.g. the discovery of scientific heuristics (Lenat, 1982), but it is not expected that they will replace or fully comprehend human creative talent.

Also consistent with this latter approach is the belief that the application of machine heuristics to scientific achievements of the past is anachronistic in that the heuristics are largely the products of recent research. Earlier scientists cannot be expected to have anticipated them. However, it is also meta-heuristically plausible that creators of artificial intelligence should look to human problem solving behavior for suggesting effective heuristics for machines, and such behavior is only arbitrarily limited to present science. Conversely, there may be heuristics applicable only to the well-defined problems within the capabilities of present machine intelligence that are similar in philosophically and historiographically illuminating ways to those that can be found in studying the behavior or the overt methodological acknowledgements of scientists of the past. Knowledge of machine heuristics may suggest plausible procedures for certain kinds of problem in epistemic contexts of the past, and thus it may guide us in what to look for in or what to infer from the often only suggestive statements and practices of past scientists.

If heuristics are not expected to be universal but are to be prospectively useful, it must be possible to choose them for a given set of circumstances in advance of their successful application. Thus there should be general classifications of heuristics and of epistemic circumstances in which they might be applied and also rules or guidelines specifying which kind of heuristic is appropriate to which kind of circumstance. Such rules or guidelines should be part of a methodology of heuristic choice, and this methodology would be an important part of a projected logic of discovery.

A heuristic or an ordered sequence of heuristics may constitute a path of inquiry, and the paths of inquiry reported by historians may consist of such a sequence of heuristics. Thus paths of inquiry in the history of science need not be sequences of historical accidents or methodologically uninteresting causal processes producing scientific discoveries. They can be, though they are not necessarily, concatenation of actions rationally chosen for their appropriateness to a rationally chosen goal. Heuristics thus can be sought in the history of science, but when a path of inquiry is drawn from history some argument beyond its occurrence or its concatenations in an important discovery episode must be offered to establish its importance as an epistemically normative guideline for scientific research.

That psychologists might study creative processes does not exclude the possibility of normative guidelines for these processes. Psychological studies by introspection or by interviewing inquirers could turn up common procedures of search or conceptualization which can be given normative or heuristic import on grounds of their apparent frequency of success (Hadamard, 1909, Gruber, 1974, Holland, *et. al.*, 1988), and possibly other grounds as well. Their presence and value might be accounted for by a process of natural selection from variant sequences of actions. This kind of empirical study can also be undertaken by historians, who could extend its scope to the work of past inquirers. But psychological studies sometimes focus on the occasion in which a novel idea occurred to the exclusion of the development of the problem solved by the idea, or to the exclusion of procedures, such as comparisons with known solutions to similar problems, that are undertaken as guidelines for investigating the problem. But 'solutions' lack scientific or epistemological significance apart from consideration of the problem they solve, and hence ignorance of the problem's justification and perhaps also aspects of its history would be ignorance of essential attributes of the discovery episode. Also psychologists need not consider the possibility that guidelines themselves can be normatively appraised in terms of estimates of their efficacy in producing solutions to problems. However, these normative considerations can be regarded as additions to or sometimes conclusions from psychological studies about how new ideas initially occur, not as disputing such psychological conclusions.

To sum up, the discovery process has these components: (i) Problems are discovered, generated, formulated and reformulated, and they are chosen or appraised for their importance, that is, they are ordered in importance for inquirers whose resources are limited. Problems embody epistemic and pragmatic desiderata, that is, they consist of items of possible knowledge and possible actions not possessed by or not within the current achievements of a community of researchers but are desired by that community. There may be normative procedures of seeking, specifying formulating and appraising problems, and these may be considered candidates for at least a portion of a 'logic' of scientific discovery. Some of these procedures may be founded in 'logic' in a more narrow sense, that is, in the sense of abstract, universal and apodictic normative principles for deductive and inductive reasoning, but others may not have such foundations. It is also debatable whether any such principles exist for non-demonstrative or inductive reasoning, and some philosophers (Mill, Quine, Putnam, Lakatos) have even gone so far as to challenge the existence of such principles even for deduction. Hence, for our purposes 'logic' will be used to designate normative procedures whether or not they are contingent or specific to a subject-matter. (ii) Problems are pursued, that is, their solutions are sought. There may also be normative guidelines that facilitate or make possible this pursuit, as heuristics can in some cases prevent an exponential explosion of possible paths of inquiry placing solutions out of reach of the fastest computers as well as human intelligence. Heuristics are normative but are generally acknowledged to be contingent and not universal in their applicability, and as normative procedures they too can fall under the heading 'logic' a broad sense. Also, since heuristics are contingent, their appraisal is a problem, and there may be normative procedures for the appraisal of heuristics. Again these can be accommodated under the broad 'logic' category.

The various normative procedures just described constitute the most plausible candidates for a 'logic of discovery', and their exemplification and specification will be the aim this book. It is not expected at this point that these principles can be usefully formalized or axiomatized because their locality, their applicability only to various limited domains, makes it likely that few significant deductive relations will be found among them. Rather, it is more likely that these normative principles will be like empirical truths in that they and their effectiveness will be matters of *a posteriori* discovery. Science is thus construed as a process in which not only do scientists learn about their subject matters, but they also learn how better to inquire as science progresses.

1.3. THE METHOD OF HYPOTHESIS

The arguments against the methodologies of constructive empiricists and inductivists purport to show (i) that there are no general 'logical' procedures for defining or constructing significant 'theoretical' scientific concepts, such as concepts of electrons and other micro-objects, of properties such as isotopic spin, of evolution or other long term and slowly acting biological and geological processes, or of energy, entropy or other abstract concepts. Such procedures were sought by Carnap early in his career (Carnap, 1928) and later by Goodman (1950) and have generally been acknowledged as impossible on both phenomenalistic and physicalistic bases. (ii) They also purport to show that enumerative and eliminative inductive routines cannot generate new concepts beyond those given in observational descriptions or in a list of alternatives from which some are eliminated. Many proponents of such arguments, e.g. Einstein (1954) Hempel (1966) and Popper (1958) advocate the view that science proceeds by free formulation of hypothesis from which empirical tests can be deduced or induced.

In spite of Popper's denials that there is a procedure for generating new ideas (1958, p. 31), his title *The Logic of Scientific Discovery* can still be understood without internal contradiction if 'discovery' is taken as inquiry, or the search for knowledge. Popper is not opposed to normative guidelines for inquiry, but rather, like Feyerabend later, he opposes the logical restraints to conceptualization imposed by inductivists and the methods of concept formation of constructive empiricists. In short, Popper's logic of discovery lacks any heuristic for concept formation beyond 'anything goes'. Falsifiable hypotheses are required for the 'scientific' component of human knowledge, but not for the meaningful 'metaphysical' component. Metaphysics is supposed to provide guidelines for excluding 'silly' ideas (1983), but without restrictions on what goes into metaphysics it seems that Popper would impose no guidelines upon concept formation. Still Popper's methodology contains broadly general heuristics for identifying relevant evidence and for trimming the number of conjectures. These heuristics include the deduction of observable consequences from the conjectures and the search for conditions in nature that will contradict these consequences or will confront the hypotheses with severe tests. One possible line of reasoning behind these strategies comes from engineering: To test the robustness of some device one subjects it to stresses that will make it most likely to fail, even though these stresses do not normally occur in the use of the device. If it survives these stresses, we should have more confidence in it in normal circumstances than we would otherwise have.

Hanson (1961a,b) noted that 'hypothetical' or 'hypothetico-deductive' procedures presuppose but do not generate new concepts or hypotheses to be tested. At best they are methods for eliminating or possibly confirming hypotheses whose concepts and formulation are already at hand. Hence the logical empiricist and Popperian focus on the epistemology of verification and falsification, and they avoid any methods of concept formation. However, beyond not providing any account for the formation of scientific concepts, hypothetico-deductivism has both heuristic and logical problems: The freedom it allows for the introduction of novel ideas is too much license. As we have mentioned, Popper himself is concerned with its permission of 'silly' conjectures. Relevant evidence, even if (sometimes mistakenly) thought to be easily discovered, may be expensive to produce, or at least scientists should not waste time with testing consequences of silly conjectures. Thus there must be some kind of heuristic for testing, and a plausible one is the initial selection of worthwhile ideas and hypotheses to be tested. Also it has long been well known that true conclusions, including those verified by observation, can be consequences of false premises or hypotheses. Descartes noted that the movements of the observable face of a clock can be generated by an indefinite number of different hypothetical mechanisms (Laudan, 1981). Accordingly this procedure provides no heuristic or guidelines for generating or inferring any particular theory from fact, whether or not novel conceptualizations are thereby introduced.

There have been a variety of attempts to overcome this 'underdetermination' problem. One approach is to demand that in addition to being sufficient for observed phenomena, acceptable hypotheses must be simple or the simplest alternative, in some sense of 'simplicity' (Maxwell, N. 1974). The search for a simple hypothesis could be considered a heuristic if 'simplicity' could be plausibly defined, for the rule demanding simplicity seemingly narrows the number of possible hypotheses entailing the facts to be explained. But, unfortunately, attempts to clarify 'simplicity' have led to no defensible criterion and have produced no decisive reason for the preference of simpler hypotheses.

A second approach to reducing the underdetermination is to accept only those hypotheses which provide the best explanation of the phenomena that they imply, i.e. to favor those hypothetical arguments which are 'abductive' (Peirce 1960, Hanson, 1961a,b). This procedure has heuristic merit provided some conception of explanatory power is available that can be used to select from the manifold possible conjectures implying the explanandum.

A third is to apply probability theory and to spell out by Bayes' theorem conditions under which entailed or probabilified evidence will enhance the probability of a hypothesis (Salmon, 1966). Salmon elaborates preconditions that must be met for such a procedure to give a hypothesis credibility: These include choosing hypotheses that make the evidence at hand most likely, several methods for attributing prior probability to the hypothesis and methods for excluding rival hypotheses.

A fourth strategy is to add further 'inductive' requirements to hypotheses, e.g. that such hypotheses be established as 'true causes' in some observable domain and extended to unobservable domains if certain qualifications are met, as advocated by Newton in his 'Rules for Reasoning in Philosophy' (1962, Book III). A somewhat weaker requirement than Newton's is that there be demonstrable analogies between the observable and unobservable domain to which the hypothesis is applied (Herschel, 1831).

Finally some philosophers (Hesse, 1974, Glymour, 1980, Nickles, 1985) have suggested that there are circumstances in which at least instances of hypotheses can be deduced from premises describing observations, experiments and empirical laws, as Newton was able to deduce the inverse square law of gravity from Kepler's first and third laws in the *Principia*. These circumstances include the acceptance of a number of substantive laws or principles, in Newton's case the first and second laws of motion. Similarly Darwin offered a deduction of natural selection from biological superfecundity and genetic variation. An aim of scientific research, then, is to create circumstances for such 'discoverability' of new laws, concepts and principles (Nickles, 1985), that is, to search for laws, etc. that would allow deductions of the desired principles. This kind of relation between observation and theory is different from that supposed in the hypothetico-deductive method, and its admission might even be considered a rejection of the hypothetico-deductivism of Popper and Hempel. It also provides the most severe possible heuristic constraint on theory, for it implies that just this theory is compatible with that observational data.

Each of these additional constraints on hypotheses has been offered by various philosophers as candidates for a logic of discovery, or a means of generating or facilitating in some way the generation of hypotheses bearing upon given observations. A general comment on this approach is that it takes relevant observed phenomena, phenomena that in some sense require explanation, as given. It does not attempt an account of how these phenomena are picked, from the vast world of objects, facts, relations and processes in the perceived world, as somehow interesting for science and in need of explanation. Also not all explananda are 'phenomena' in the sense of observed facts, entities, events or processes. Among those which are unobserved but in need of explanation are events in the past history of the cosmos, the Earth or the biosphere such as the accretion of localized matter into galaxies, stars and planetary systems, the most recent separation of the terrestrial continents, or the colonization of the Galapagos Islands by turtles. Others are not phenomena in the sense of events observed after occurring naturally, e.g. the properties of elemental sodium or potassium, voltaic electricity, or induction coils. These must be and must have been produced artificially before being observed.

Problems with each of these strategies for 'discovering' hypotheses will now be discussed in greater detail:

1. The metascientific concept of 'simplicity' is notoriously vague, and attempts to clarify it produce ambiguous and sometimes dubious results.

First of all, terminological and propositional economy, viz. a minimum number of undefined terms and unproved postulates, can be achieved by superficial and trivial logical operations, such as conjunction and arbitrary definition. Semantic methods for establishing terminological and postulational independence can be used to exclude these trivial operations, but they are decisively applicable only to formalized theories. But formalization does not realistically represent the epistemic circumstances in which the theoretical researcher finds himself. In some of these latter circumstances suitable concepts are sought after, not given. However, there may be some analogue of semantic independence that guides the introduction of some novel concepts, e.g. the concept of mass is needed to distinguish descriptively, if not to explain, the different accelerations produced by impulse and other non inertial mechanical forces. However, this is a strategy for increasing terminology and not for selecting terminologically simple theories, though it might be claimed that if we adhered strictly to this as a strategy for introducing terminology, redundant terminology could be avoided and terminological simplicity could be achieved. However, in the light of the apparent multiplicity of scientific heuristics, at this point we should not be prepared to acknowledge just one heuristic for introducing novel terminology. In other cases the desideratum of terminological simplicity has been violated: As Popper (1958) points out, an ellipse requires more parameters to specify than a circle, but this feature did not, and should not have kept Kepler from advocating his 'First Law' of planetary motion. Van Der Waal's equation for gases introduces terms not appearing in 'ideal' gas theories, but the behavior of actual gases require these extra terms (Harre, 1972). In other instances simplicity in some sense seems to have borne epistemic fruit: Copernicus could not have chosen heliocentricism on grounds of empirical considerations, for his heliocentric models could not be made any more empirically adequate than Ptolemy's or other geocentric models. Instead he chose heliocentricism because it eliminated accidental coincidences between the motion of the Sun and the epicycles and deferents of the major and minor planets, respectively. Einstein was so sure of his theory of gravity that he claimed that if it were not verified empirically he would consider God to have made a grave error. However, in searching for similar reductions of the other forces of nature Einstein was gravely disappointed.

Some concepts of 'simplicity' emerge more sharply with closer attention to what philosophers and scientists mean in advocating this as a metascientific desideratum. In his first 'Rule of Reasoning for Philosophy' (Thayer and Randall, 1953, p. 3, 1962, Book III) Newton stipulates that causes should be just sufficient for their effects, as well as 'true', because 'Nature is pleased with simplicity and affects not the pomp of superfluous causes'. But then, what count as 'superfluous causes'. Newton's argument for his second rule, which prescribes the same cause for the same effect, offers one possible answer to this question: We should avoid assigning different kinds of causes to effects that are the same, e.g. gravitation to terrestrial accelerations and something else to celestial accelerations, because Nature is always 'in conformity with herself'. Here the epistemic import of this kind of simplicity is based upon a presupposition of cosmic uniformity. But again, 'same effect' is vague. In the case of gravitation Newton argues that there is continuity linking celestial and terrestrial acceleration by imagining a projectile approaching the orbit of an imagined satellite that approaches the terrestrial region (1962, Book III, p. 409). Though Newton's appeals to such continuity makes a plausible case for the identification of terrestrial and celestial gravity, and thus provides a sufficient condition for searching for the same cause, it has yet to be established that this kind of continuity is necessary for the 'sameness' of effect that warrants the search for like causes. Also the 'just sufficiency' prescribed here could mean other things as well, e.g. it could exclude superfluous causes, such as Copernicus' exclusion of multiple epicycles for planetary retrogressions in favor of one terrestrial motion. In this case Copernicus may have appealed to the plausible principle that it is unlikely that there occurs in nature a large number of coincidences, as in the periods of the major planets' epicycles, the periods of the minor planets' deferences and the Sun's period, without some common cause.

An additional problem is raised by Newton's seemingly *a priori* assumption that the cosmos is simple in the senses described above. In some respects, or in some senses of simplicity, this assumption is false. The case of Kepler's ellipse and the complexity of van der Waal's gas laws are among the many good examples of nature's falling short of simplistic human concepts and ideals. Thus the assumption that nature is 'simple' cannot be made a *priori*. If the assumption were *a posteriori* it would have to be established either on the basis of direct observation, which would not be possible in the case of universal gravitation and many other aspects of nature, or as an induction from theories that are credible on other grounds. In this latter case simplicity cannot be used as a metascientific desideratum in appraising the theories to which appeal is made in establishing it as a metascientific desideratum. However it is possible to establish by historical survey that theories that are credible on other grounds happen also to be simple in some sense. Once thus established *a posteriori* simplicity could be extended as a desideratum to other theories. A similar procedure could be used to establish or refute other metascientific desiderata (Cf. Laudan's 'Reticulated Model' for establishing methods and values in science, 1984, especially Chapter 3).

One further possibility is that 'simplicity' is desirable only as a means to intelligibility on the part of human inquirers or as a means to the convenience of human calculators and predictors. 'Intelligibility' is here taken to mean accessibility to human understanding in the sense that humans can grasp and manipulate the ideas with desired simplicity so that their application might be extended in the domain of inquiry, or so that novel phenomena in that domain might be predicted for the purpose of extending empirical knowledge of the domain. Such activities are instrumental to testing the theory, or for other practical purposes serving human needs. However, in this case the desideratum is pragmatic simplicity, not epistemic simplicity, where epistemic simplicity is proposed as a criterion for the credibility or truth of the simple theory. Pragmatic simplicity is also as vague as is our conception of human intellectual capabilities, though doubtless it can be attributed to specific humans in regard to specific theories and conceptual systems. But such attribution leaves unanswered the question of whether simplicity has any epistemic import, i.e. whether it can be taken as a criterion of truth.

2. In regard to 'abductive' procedures, the principal philosophical problem is to determine what constitutes the 'best' explanation. To argue that the best explanation is that which is the best predictor of novel phenomena or of novel phenomena that can be established in fact begs the question raised by the critic of hypothetico-deductivism, viz. how are we warranted in inferring the truth of premises from the truth of conclusions or in inferring the truth of antecedents from that of consequences. To argue that the best explanation is the simplest explanation is to raise again the problem of epistemic simplicity. To argue that abductive explanation is inference to the only available explanation raises questions about the circumstances in which this uniqueness occurs: The only avail-

able explanation may be the only one that inquirers can think of at a given time, but these circumstances may just be indications of the conceptual limitations of the inquirers, not indications of the way the world is structured.

Nevertheless Darwin and many other scientists and philosophers have used abductive arguments in defense of, e.g. evolutionary hypotheses and realistic interpretations of scientific theories. Also Peirce, Hanson and others have argued that abductive arguments are particularly important in scientific research as a means of establishing which proposals are initially worthy of developing into testable theories and worthy of the effort of searching for, designing and producing empirical tests. They offer a solution to what we have described as the heuristic problem for hypothetico-deductivism. Hence criteria for 'intelligibility' or explanatory import, possibly other than the usual senses of 'simplicity', and their heuristic and epistemic implications might still be sought. Peirce gives this account of abduction:

> ... abductive inference shades into perceptual judgment without any sharp line of demarcation between them; or, in other words, our first premises, the perceptual judgments, are to be regarded as an extreme case of abductive inferences, from which they differ in being absolutely beyond criticism. The abductive suggestion comes to us like a flash. It is an act of *insight*, although extremely fallible insight. It is true that different elements of the hypothesis were in our minds before; but it is the idea of putting together what we had never before dreamed of putting together which flashes the new suggestion before our contemplation. (Peirce, 1960, Vol. V, Section 181)

According to Peirce, abductive insight has these characters: (a) it is sudden and uncontrolled as is normal perception (possibly an affirmation of the illuminist's views of discovery described above),

(b) what little evidence the insights have is apparently intrinsic, i.e. in a sense they are self-evident, (c) it can assemble elements that have occurred to us before in new ways that have not previously occurred (cf. heuristics of conceptual combination in Chapter 3 below), and (d) they are integrative in that they relate disparate elements generally by means of internal relations (cf. the discussion of Hanson's gestalt shift analogies immediately below). Presumably what differentiates visual insight from abductive is that abductive insights need not be limited to visual space or visual properties and relations, but can be in terms and dimensions somehow apprehended by the mind.

In arguing that abduction is essential to the acquisition of knowledge, Peirce calls attention to the heuristic value of abduction:

> ...possible theories, if not strictly innumerable, at any rate exceed a trillion— ...and therefore the chances are too overwhelmingly against the single true theory in the twenty or thirty thousand years during which man has been a thinking animal, ever coming into any man's head. (Peirce, 1960, Vol. V, p. 238)

Abduction thus assists the researcher by recommending the narrowing of possible hypotheses for further investigation. Peirce attributes to abduction the greatest possible heuristic power: the best explanation is *the* natural hypothesis which fits the mind's pre-adaptation to nature. This move would seem to solve the underdetermination problem, though it remains to be seen whether 'natural' hypotheses are really unique, whether they are adequate to all problem situations that come up in science and whether their naturalness is any reason whatever for presuming they might be true.

Peirce attempts several different explanations of what he takes to be the abductive faculty. He tells us that a chicken does not, upon hatching, need to rummage through possible acts before picking up a grain. Humans should not be denied what a chicken possesses, *viz.* "...an innate tendency toward positive truth...". Thus "...man's mind has a natural adaptation to imagining correct theories of some kinds, ... without some glimmer of which he could not form social ties and consequently reproduce his kind". (Peirce, 1960, Vol. V. p. 238). These innate capacities are not just for appreciating causes and explanations of human actions, but also for explanations of mechanical and other physical occurrences.

Peirce's account of abduction even has a hint of Platonism: "What can our first acquaintance with an inference, when it is not yet adopted, be but a perception of the world of ideas?" (Peirce, 1960, Vol. V. p. 192). He also suggests that the abductive faculty can be explained as a product of evolution:

> ... if the universe conforms, with any approach to accuracy, to certain highly pervasive laws, and if man's mind has been developed under the influence of these laws, it is to be expected that he should have a *natural light*, or *light of nature*, or *instinctive insight*, or genius, tending to make him guess those laws aright, or nearly aright. (Peirce, [Tomas, ed. 1957], p. 245)

The supposed affinity of the human mind for the laws of the universe or for Platonic archetypal forms is subjectively evident in something resembling Aristotle's intuitive induction: "... it must be good reasoning to say that a given hypothesis is good, as a hypothesis, because it is a natural one, or one readily embraced by the human mind" (Peirce, 1957, p. 238). The 'naturalness' of a hypothesis is perceived by this 'natural light' or by its intuitive intelligibility.

Peirce's 'abduction' is not unlike Aristotle's 'induction', 'the method by which sense-impression implants the universal'. Like Pierce, Aristotle compares induction to sense perception, which, with memory, are natural faculties in the human being as well as in some animals. Elements appearing in repeated sense perception and memory can fall into a pattern of experience, as a retreating army falls into a formation one soldier at a time. The mind thus has a capacity for perceiving the universal in the particular, where sense perception is of the particular but its content, certain patterns or forms, is universal. (*Posterior Analytics,* (1941) Book II, Section 19)

Peirce thus claims that humans are endowed with an instinctive mechanical intuition through which abductive evidence in physics can be perceived. It is through this perceptual faculty that humans perceive intuitive intelligibility. It is not clear what this intuitive intelligibility actually comes to. If it is a kind of visual picturability it cannot be used as a necessary condition for scientific explanations because many of the accepted explanatory theories today are not readily picturable in visual terms. Peirce's recognition of this point might be implicit in the distinction he draws between abduction and sensory perception and in his suggestion that abduction is perception in the 'world of ideas'. Functions of more than three spatial and one temporal variable and functions with imaginary dimensions that are often used in physics can be partially represented in visual space by means of two dimensional graphs, but the mathematical structures in greater than three dimensions or in imaginary dimensions cannot be regarded as visual or accessible to visual imagination. What features beyond the properties and relations accessible to sense perception would make a hypothesis 'natural' in Peirce's sense is not clear.

Other philosophers, after reflection upon counterintuitive aspects of contemporary physics, would probably reject Peirce's belief that explanatory import is intuitively manifest, though it is unclear how important such intuitiveness is to Peirce. Intuitive intelligibility in the sense of visual picturability can no longer be regarded as an obligatory explanatory value because many acceptable explanatory theories, e.g. quantum mechanics and quantum electrodynamics, do not fulfill it.

Hanson (1961a) puts in place of Peirce's intuitive intelligibility certain 'logical' properties that all explanations should have:

> 1. An explanatory hypothesis should be expressed as a mathematical relation between two parameters.

Mathematical functions such as that determining the behavior of the force of gravity in space are the preferred form of explanatory hypothesis, according to Hanson.

> 2. Explanatory hypotheses should link parameters into 'patterns', as the gestalt picture is composed of a pattern of internal or organic relations.

Patterns are a complex of causal relations and relations of ontic and other dependency. Of course the patterns in gestalt pictures are visual, but these visual patterns have certain characteristics that may be abstracted from visual space and applied in non-visual spaces. It is these abstract characteristics that may bear what is of importance in intelligible or explanatory theories.

If the relations are internal, they take part in defining or conceptualizing the elements related. In gestalt pictures the relations are internal: A line represents the jaw of the *jeune fille* in the Toulouse-Lautrec (Hanson, 1961a, p. 11) gestalt picture just if it occurs in a certain context. In an application to physics we might organize the possible paths of a photon in terms of causal processes whereby the photon is propagated in space and time with phase changes whose rate depends upon the photon's energy. The phase changes determine whether superimposed photons cancel or reinforce in one or another region to which propagation is possible. Opaque barriers, partially reflecting surfaces, sources and detectors are all part of a network of causal and spatial relations that describe process of radiation in commonplace circumstances. An opaque barrier in this context is defined as something that prevents the penetration of photons by absorbing or reflecting them. The presence of such barriers influences the possible trajectories of electrons, where possible trajectories are needed for calculating probabilities of various states of the system in which the electrons occur and move. Some of the items in this arrangement have particular effects on the phase of the photons interacting with them. A fully reflecting surface, for example, has in this context, beyond a variety of other attributes commonly ascribed to such, the property of reflecting a photon with a probability of one and changing its phase 180 degrees. Photons and their properties are essentially involved in describing reflecting surfaces and other components of optical processes.

Hanson's condition 1 can hardly be accepted as a requirement for all good explanations. Biological science is full of good explanations that make no use of mathematical functions. However, Hanson adds two further requirements:

> 3. Explanations must not be circular, that is they must be in terms not used essentially in the description of the phenomena explained.
>
> 4. Pattern statements must be unrivaled in their patterning.

Condition 3 seems plausible enough and needs no further discussion. According to condition 4, the pattern must be in some superior to rival patterns to be credible. This is the uniqueness condition in abduction, that is, reasoning to *the best* or *the only* explanation.

Various criteria are often cited as counting for superior explanatory power, but they are not made entirely clear by either Hanson or Peirce. Some such criteria are as follows: Consistency with background beliefs (where what we believe is what we accept as true) is a necessary condition for accepting as true explanatory as well as other hypotheses. Also internal coherence is a value cited by Copernicus as lacking in geocentric planetary theories, and by this he meant that geocentricists used various combinations of central and peripheral epicycles, deferents, eccentric locations and equants in their models of the Moon, Sun and five planets. Accordingly geocentric theories of planetary motion were less than ideally internally coherent. Darwin claims that 'simplicity' and 'sublimity' are lacking in creationist theories, and by this he means that the creationist theories need not consider biogeographical circumstances beyond the benevolence of God in giving an organism what it needs in an environment in their explanations of biogeographical distribution (DeBeer, 1960, B101f). In creationism the proximity of habitat in space and time has nothing to do with the similarities of the animals occupying those habitats. However, evolutionary explanations causally relate these features of habitats to similar species and genera occupying these habitats. Internal coherence is also sought by Einstein in his hoped for unified field theories: Rather than construing forces as extraneous relations between masses, charges and magnetic poles, Einstein sought to construe them as internal characteristics of space. Internal coherence can be construed as the relative absence of *ad hoc* components in a system of concepts and laws that are applied under various circumstances to generate explanatory models for happenings in those circumstances. The ability to accommodate in the explanatory network the features of other like physical arrangements, that is the general applicability of an explanatory scheme, is another metascientific desideratum relevant to the comparative ranking of such schemes in terms of their explanatory power (Maxwell, N., 1974, Laudan, 1977).

According to Hanson, conceptual systems meeting the above four conditions have special epistemic status:

> Pattern statements are not falsifiable in the same sense as detail statements. ... A pattern statement's ... negation does not represent the same conceptual possibility, for it concerns not observational detail but the very pattern which makes those details intelligible. (Hanson, 1961a, p. 87)

It is this special status which gives the abductive inference its epistemic merit as an inference to credible propositions. However, it has not been shown that such coherence is an indicator of the truth of the patterning hypothesis. How can we be sure *a priori* that the world is as coherent as is our patterned conception of it?

Peirce offers an *a priori* reason to assume that explanatory schemata having these coherence and applicability properties do represent nature: The mind is adapted by evolution to being predisposed to representing nature as it is. Such adaptation might be expected for those aspects of the world encountered in everyday experience, whether it concern the behavior of other persons or animals, the natural environment, or the behavior of mechanical and other physical contrivances. The predispositions of the mind are plausibly particularly adapted to survival needs. Uniformity in nature is one such need: Organisms can exploit uniformities to their advantages by learning from sensory experience and remembering what helps and hurts. Such learning utilizes the natural capacities mentioned by Aristotle. Assuming uniformity, these perceptions give the organism advantages in the future not possessed by organisms not capable of sensing, remembering and adjusting behavior on the basis of presumed uniformities. If we don't expect nature to remain uniform there is no unique specific alternative expectation warranted by past experience. It might change uniformly at any of an infinite number of rates, or it might change non-uniformly in an infinite number of ways. This perceptual capacity is no guarantee that nature will be uniform in the required sense, but it does allow the organism to take advantage of whatever uniformities might occur.

However, can we expect such adaptation to extend to the more esoteric world of molecular and atomic physics that is available to us only through highly developed technologies of experiment and observation? Then, perhaps, it is our intellectual adaptability to the esoteric that has survival value in the evolutionary process, not some habit of repeating what has proven successful in the past. Should we expect to be endowed with capacities to adapt in entirely new ways to esoteric worlds as opposed to, or perhaps in addition to, our capacities to anticipate the more ordinary world of daily experience? Perhaps these capacities to adapt intellectually can be given an evolutionary justification in the sense that those having them are, in appropriate circumstances such as a world that changes in unanticipated ways, fitter than those without them. But then we cannot expect success in anticipating *a priori* what happens in such esoteric worlds. In such exotic worlds it would seem that abduction has no epistemic force whatever. On what basis might we expect some entirely novel kind of explanatory network to represent the world just because it has intrinsic coherence relations, e.g. it contains many fewer arbitrary or *ad hoc* elements? Is it that our intellectual endowments and adaptability are so limited that such esoteric worlds are intelligible to us only in so far as they are similar to those more familiar?

In this case some *a priori* correspondence between concept and even exotic world can be explained as a product of evolutionary adaptation. Only then we can argue abductively to novel conceptual schemes which are in some important respects like those that have proven successful in everyday life, and prior expectation of truth is justified because of evolutionary adaptation of thought to world.

The second variant of Hanson's abductive procedure (Hanson, 1961b) seems to favor the non a prioristic conception of abduction, viz. that of an intrinsically adaptive mind that adjusts to the world in the process of learning and then extends the patterns acquired in this adjustment to further domains of the environment. The second variant of Hanson's abductive argument takes this form: H is the *kind* of hypothesis that has successfully explained phenomena like P in the past. H explains P, so H may be true. Adapting from past experience and not being genetically adapted seems contrary to what Peirce had in mind but also gives human inquirers greater flexibility in dealing with exotic worlds. In a highly varied world such adaptive behavior seems to have advantages over genetically programmed behavior. Both preprogrammed and adaptable intellects have similar long run advantages in worlds that are not so varied. But in worlds that vary, adaptive intellects have an advantage, so it would seem that adaptive intellects have an overall advantage. Since we are not sure which kind of world prevails, it would seem, contrary to Peirce, that adaptability is generally fitter. (See also Kantorovitch and Ne'eman, 1989, p. 506.)

3. Turning now to another approach to bolstering the hypothetico-deductive method, Bayes' theorem is as follows:

$$P(h/e'\&e) = P(h/e) \times P(e'/h\&e)/P(e'/e)$$

According to an application of Bayes' theorem that a number of philosophers are still discussing, the probability which a given piece of empirical evidence e' gives a hypothesis h, h's *posterior probability* is directly proportional to its *prior probability* viz. the degree to which h is confirmed by empirical evidence e obtained earlier, and to the *likelihood* of h, viz. the degree to which e' is expected given that h and prior evidence e is true. In an extreme case this last value is unity, that is where h & e logically entails e'.

Salmon (1966) points out that this account is more general than traditional accounts of hypothetico-deductive procedures in allowing for the likelihood component that h can weakly entail e or render e probable rather than entail it. Also the confirmation of h on the basis of e' is inversely proportional to the dependence of subsequent evidence e' on prior evidence e, or, equivalently, inversely proportional to the degree to which e' is expected on the basis of hypotheses h', h'',...etc. that are exclusive alternatives to h.

These relations do capture certain methodological intuitions occurring among scientifically minded individuals, including philosophers such as William Whewell who addressed the problem of hypotheses in science in the nineteenth century. That h's posterior probability is directly proportional to its prior probability captures the intuition that the credibility of a hypothesis can increase as positive evidence for it accumulates. That it depends directly on the likelihood of h captures the intuition that relevant evidence for h is the evidence that is probable if h is true, though intuition also tells us that this circumstance is insufficient to add to the credibility of h. The evidence must also be improbable if h is not true, or if something incompatible with h is true.

It is highly probable that I'll be late to work if my bicycle tire explodes on the way in, but my being late is not generally evidence for such an event, for lateness could equally be attributed to a failure of the alarm clock, snowy conditions, a slow leak in the tire, etc. For lateness to be evidence for an exploded tire there must be some reason to believe that lateness is exclusively probable on the grounds that the tire exploded, and these would occur if it were believed that slow leaks, snowy conditions, or faulty alarm clocks provide no hindrance to the process of arriving at work punctually. Thus the probability of the exploding tire is inversely proportional to the degree to which lateness is expected on the basis of alternative hypotheses, including slow leaks, etc., all of which can be taken as excluding the explosion of the tire.

A third intuition is satisfied by an equivalent formulation of Bayes' principle: The confirmation provided h by e' is inversely as the degree of dependence of posterior evidence e' on prior evidence e. This point captures Whewell's intuition that the best 'inductions' 'colligate facts' and are 'consilient', viz. they are supported by distinct facts and classes of facts, or facts whose evidences are mutually independent.

However, beyond the intuitive plausibility of applying Bayes' theorem to hypothetical induction, there remain difficulties, particularly in interpreting the probabilities attributed to hypotheses, empirical evidence and conditionals connecting these. Although the probabilities of events, such as the occurrence of double-ones in one or several tosses of presumably fair dice, are uncontroversial it is not equally obvious that probability can be assigned to propositions or statements in any exact and epistemologically illuminating way. In the case of exploding bicycle tires, and other engineering projects, it is commonplace to statistically rate the reliability of tires, cables, spokes and the like by means of tests or simulated use and the frequency of failure thus observed is projected into circumstances of actual use. These can be a source of prior probabilities in the reasoning cited above. Similarly a person can appeal to his past experience in a certain style of commuting to estimate the frequency of various disruptions of punctuality. However, when it comes to estimating the credibility of a scientific hypothesis, e.g. universal gravitation as in evidence from terrestrial projectiles, pendulums, and the weight of known masses, or from celestial regularities such as Kepler's laws, there appears to be no obvious link between the frequency with which gravitation and projectile motion occur together and some estimate of the probability that gravitation will be associated with observed projectile trajectories. Gravitational fields are not observable apart from terrestrial and planetary trajectories or the weights of bodies of known mass, and hence the kind of inductive procedures used by engineers for testing mechanical devices are inappropriate for theories of gravity and other physical theories that appeal to hypothetical entities, processes or forces. A fundamental consideration in philosophers' defense of hypothetico-deductivism is that some ground be given for assigning credibility to hypotheses about unobservables.

A second possible interpretation of epistemic 'probabilities' is that they are measures of degree of warranted belief. The accumulation of positive evidence e, e',... for h enhances the degree to which an individual is warranted in believing h, and the likelihood of h is the degree to which he or she is warranted in expecting a piece of evidence given that h is true, or in believing that there is a kind of conditional relation between h and e, and the dependence of e and e' is the degree to which he or she is warranted in expecting e' given that e is true.

However, since degree of warranted belief is needed in the prior probabilities, estimates of independence of evidence, and likelihoods of hypotheses before posterior warranted belief can be calculated, Bayes' theorem presupposes but does not generate knowledge of what constitutes warranted belief or what circumstances warrant belief. Hence, as in the case of frequency interpretations of 'probability', Bayes' formula does not provide a general account of how or on what grounds epistemic probability can be attributed to any belief, particularly theoretical claims such as Newton's law of gravity or the theory of evolution by natural selection.

Also, since all empirical investigations must begin with obtaining initial empirical evidence, viz. some e' before which no further e has been obtained, the prior probability of h must in actual investigations be estimated initially independently of empirical data. Salmon (1966) has suggested that these prior non-empirical estimates be based upon epistemic criteria that constitute a logic of discovery, which he understands as a rationale for initially appraising hypotheses and theories as worthy of the investment needed for further empirical investigation. He cites as possible non-empirical criteria simplicity, the authority of experts and compatibility and coherence with accepted background theory. But Salmon's position underscores what lies beyond epistemological applications of Bayes' theorem: It is insufficient to decide the epistemic authority of such considerations as simplicity, expert authority and external coherence, much less the relevance of certain specific kinds of observable or observed information, such as the relevance of observed to actual motion, of fossils to prehistoric past entities or events, of vapor trails in supersaturated water vapor to charged sub-atomic particles, or of pointer readings to amperage. The pursuit of such evidential relations is an essential part of science for which philosophy and mathematical probability theory are insufficient. It is only in the knowledge of these relations that scientists can obtain estimates of the reliability of their hypotheses about the past or about entities or processes that are not observable because of their size, remoteness in time or space, or their slowness.

In sum, Bayes' theorem does little more than systematize a few widely affirmed methodological intuitions and never touches the truly interesting epistemological problems such as assessing the epistemic import of putative non-empirical epistemic criteria (see the discussion of simplicity and abduction above) and discovering means of observational or experimental access to hypothetical entities, events and processes. It is doubtful that epistemic applications of probability concepts are more than metaphorical, and it is equally doubtful that such metaphors are the kind that can generate genuine insight, in this case into epistemological issues. Thus applied probability theory is not very promising as a scientifically useful means of strengthening deductively invalid hypothetical inferences to the point of being heuristics for discovery.

4. Newton's *vera causa* procedure, known to some as *transduction* (McGuire, 1970) can be interpreted with various degrees of strictness, as indicated in the above discussion of simplicity. On the one hand one might require that the 'same' cause be exclusively cited for what counts as the 'same' effect. There is vagueness in 'same cause' and 'same effect', but implicit in Newton's reasoning is the stipulation that centripetal accelerations in any body constitute same effects and that applied centripetal forces are the prescribed same causes provided they can be observed in one instance of the same effects. Thus, given that terrestrial projectiles are accelerated centripetally with respect to the Earth, that Kepler's second law (the areas law) logically entails that planets move with centripetal accelerations, that his first (elliptical orbits) and third laws (mean distance is the $\frac{3}{2}$ power of period) logically entail motion under inverse square centripetal forces, and that the moon's centripetal acceleration toward the Earth can be computed

to be in agreement with observed characteristics of its orbit, the centripetal accelerations are sufficiently the same to warrant the extension of gravity, which is observed in the terrestrial region, to the celestial domain. 'Gravity' is the force of mutual attraction of two or more material bodies and is observed terrestrially as the weight of projectiles and other terrestrial bodies.

Newton's Rule III permits the inductive extension of qualities which are observed without 'intensification' or 'remission'. The qualities Newton mentions in his defense of this rule include extension, hardness, impenetrability, movability, and inertia. What is meant by 'intensification' and 'remission' is somewhat obscure, for presumably some bodies do change in size, grow soft, change in their motion or acrete or diminish in mass. It is likely that Newton is here obscurely formulating a version of Descartes' argument that extension is essential to all matter: If we remove in thought the extension, hardness or impenetrability, movability or inertia of a body, no body remains in thought. Also, importantly, these qualities are not removed by dividing bodies into components, for extended bodies are composed of extended parts, as hard or impenetrable bodies must have hard or impenetrable parts, movable bodies movable parts, and massive bodies massive parts. These qualities may be described as non-emergent in the sense that they may be ascribed to bodies and their parts without committing any fallacy of composition or division. Newton's 'same effects' may thus be stipulated to be effects describable in terms of these primary qualities. In this case, 'sameness' is used less restrictively than in the case of projectiles and planets. However 'sameness' thus interpreted can make available Newton's Rule II as a means of applying the *vera causa* method to micro-processes, such as supposed particulate components of a gas. These components can be expected to manifest the 'same' characteristics of observable bodies, viz. extension, hardness, motility, inertia, and their motions should have the 'same' causes, in this case not necessarily gravitational force, for their motions are apparently not centripetal accelerations, but possibly something else causing the centrifugal accelerations which can be shown to entail Boyle's gas law (pressure $\times$ volume $=$ constant).

It is thus that Newton achieves a form of 'discoverability' or a relatively strong inference from observed effects to causes (Nickles, 1985). This kind of procedure has been applied by other important figures in the history of science, particularly the geologist Lyell, in defending actualistic and uniformitarian geology. Also Darwin uses it with interesting variations in his search for a genetic mechanism of speciation. However, there may be situations in which scientists need to infer to hypotheses about unobservable entities or processes with no justification for the belief that the 'same' effects in either of Newton's senses occur in both the unobservable and observable domains in question. Not all qualities or other attributes of things are non-emergent as are Newton's mechanical 'primary qualities'. Also it should not be assumed *a priori* that all domains of inquiry can be fundamentally intelligibly described in these mechanical terms, as seems to have been demonstrated in the development of quantum mechanics and other terms for describing and explaining the properties and behavior of the various fundamental particles.

There remains the possibility that 'sameness' can be used less restrictively than Newton's use, as suggested by the early 19th century philosopher John Herschel (see his 1830) and by other scientists (James Clerk Maxwell) and philosophers (particularly Mary Hesse, 1962, 1966, 1974 and Rom Harre, 1970). 'Sameness' could be replaced by certain kinds of 'similarity' and thus provide a more flexible procedure than Newton's for restricting hypothetical inferences, but this possibility needs further detailed discussion because the procedure could be too flexible or permissive thereby giving no heuristic advantage over the fallacious inferences from conclusion to premises or from consequent to antecedent. This possibility will be discussed further in Chapter 5.

In sum, Nickles' 'discoverability' is not without its problems. First, deductions of interesting theoretical hypotheses from phenomena require known or assumed laws, such as Newton's law of inertia and the formulation of these laws may require the introduction of new concepts, such as Newton's concept of mass, which is in no apparent or intelligible sense deducible from observations or observational concepts. Second, even though Newton effectively sets up phenomena from which essential components of the law of gravity are deduced, particularly centripetal forces are deduced from Kepler's areas law and the inverse square property is deduced from the $\frac{3}{2}$ power law and the ellipse law, the discovery of Kepler's laws themselves and the program of explaining celestial motions by forces manifest in the terrestrial region has a history that is not obviously explainable in Nickles' terms of discoverability. Some of this history will be discussed in Chapter 2 below (see also Kleiner 1983), where Kepler's research program and his discovery of the First Law (the ellipse law) and the Second Law (the equal time equal area law) will be described. Third, there appear to be limits in the applicability of Newton's *vera causa* method, particularly to unobservable domains that cannot be assumed to be qualitatively the same as observable domains. Possibly the method could be extended to accommodate discoveries of properties, processes, laws and intelligible theories in these domains, but there remains a danger that such extensions reduce the heuristic efficacy of the method. These points will also be further discussed in Chapter 6.

As Popper, Kuhn, Feyerabend and others have argued, most scientific research is not carried out in a theoretical vacuum. Background theoretical assumptions about a domain of inquiry and about related domains are necessary in defining the domain, identifying elements within it, raising meaningful questions about it, and designing experimental and observational strategies for gaining empirical access to it. These assumptions play a heuristic role of constraining scientific inquiry: They can constrain both the epistemic objectives of inquiry, i.e. the problems chosen, and the means for achieving these objectives. Some conceptual and empirical novelties fit within this background framework, but often historically significant epistemic novelties include a change in background theory that entails a more or less profound change in conceptions of domain, problem or experimental strategies. Thus one aspect of the problem of discovery, particularly the problem of discovering an explanation of given phenomena, is the appraisal of these constraints and the direction of research seeking their revision or replacement (Nickles, 1980a,b).

Certain philosophical questions can be raised about the status of these constraints that bear upon the problem of discovery: Do they represent restrictions on possible knowledge of the world due to some all pervading features of the world or features of the interaction of inquirers with the world, e.g. as some claim about the import of the Heisenberg uncertainty principle? Or is their importance just pragmatic, that is, is their role only to assist inquirers who are limited in their resources and capacities for knowledge and action? If they are epistemic, then their justification must be epistemological and at least this aspect of the 'logic of discovery' is not distinct from the 'logic of verification' or the search for criteria of truth. On the other hand, if they are just pragmatic, then in this case the thesis that guidelines for discovery have no epistemic import is sustained. 'Merely heuristic' considerations then are distinct from epistemological items and there may be some basis for philosophers' scorn for the former.

Another issue implicit in recent philosophical discussions of discovery is how much constraint should be placed on the 'freedom' to invent hypotheses or to introduce concepts in terms of which novel hypotheses can be conceived or formulated. Philosophers of the logical empiricist persuasion have held that scientific methods are governed by the objective of formulating empirically responsible theories and this objective can be achieved by means of methods backed by deductive and inductive logics. Scientific rationality thus consists simply in adopting this aim and conforming to the principles of deductive logic and probability theory. Since these aims and procedures place no constraints on the introduction of novel concepts, there is no 'logic' of discovery and the creation of new ideas is an a-rational process. Scientific innovation consists simply in the production of new concepts by some non-logical intellectual process, but testing consists in the application of logical principles in determining observable consequences to be established or rejected. The rules of scientific procedure, for testing, verifying and falsifying scientific theories, are universal and immutable, but they do not uniquely determine nor constrain in substance the theories that succeed in accord with the logical empiricists' objectives and procedures. Neither the objectives nor the procedures guarantee the truth or the permanent success of theories that conform to them. Discoveries consist of the failure of once successful theories and the production of replacements, but they influence neither the logical empiricists' objectives for science nor their means for achieving these objectives. Following Popper, this methodology can generate refutations but it cannot generate replacement theories. In contrast, the objectives and guidelines for the conduct of science can and should be seen to transcend formal logic and mathematical probability theory, as suggested in the discussion of Baysean induction above. Toulmin (1972) has suggested that scientific rationality should be seen not as reasoning in accord with immutable logical principles, but as opportunistic strategies of gaining epistemic access to and adapting epistemically to various aspects of our surroundings. In addition to allowing the possibility of non-standard logics and non-classical probability theories, such opportunism can include strategies for employing the fruits of scientific inquiry for the extension of scientific knowledge. That is, the credible propositions achieved about a domain of inquiry can themselves be employed in the search for further knowledge of the domain. In the cases considered below it will be seen that such substantive beliefs play a far larger role in guiding inquiry than any of the traditional 'logics'. Accordingly, "logic of discovery" may be a misnomer if it suggests that something like formal logic or abstract inductive logic plays an important role in scientific inquiry. Rather "logic" should be interpreted in the title of this work much more broadly to mean guidelines for the deliberate and systematic conduct of scientific inquiry with the objective of facilitating epistemic progress.

The objective of the next chapter is to explore the role that conceptualizations have in scientific discovery. This effort is preliminary to addressing the questions just raised.

1.4. SUMMARY

Philosophers' tendency to shy away from the problem of discovery is partly due to inadequacies of certain inductive and constructive procedures for scientific concept formation. Some philosophers view these procedures as overly restrictive and as excluding the formation of certain successful conceptual systems, such as those of relativity and quantum mechanics. Historicist philosophers have been no more successful in addressing this problem than have been traditional logical empiricists, largely because the historicists view scientific revolutions as sweeping change of concept, ontology and method, leaving no guidelines for revolutionary research. Normal scientific research is guided largely by substantive beliefs regarding the subject matter and neighboring domains. Scientific innovations can include changes in these guiding principles. The philosophical problem of the rationality of scientific discovery is one of determining how it is possible to rationally challenge and replace such guidelines as well as to follow them.

CHAPTER TWO

SCIENTIFIC INQUIRY

> The properties of particles are discovered and (in a way) determined by the physicist. Phenomena are observed which are surprising and require explanation. The observations may be of the tracks left in a cloud chamber or in a photographic emulsion, or the scintillations excited when particles strike certain sensitive screens, or one of a number of other indirect effects. The theoretician seeks concepts from which he can generate explanations of the phenomena. From the properties he ascribes to atomic entities he hopes to be able to infer to what has been encountered in the laboratory: he aspires to fix the data in an intelligible conceptual pattern. When this is achieved he will know what properties fundamental entities do have; and he will have learned this be retroduction. (N.R. Hanson, 1961a, p. 123)

2.1. INTRODUCTION

The aim of this chapter is to provide an overview of the various kinds of inquiry that constitute scientific research. It is intended to encompass something similar to what Kuhn (1970) called 'normal science', but also to consider the possibility of revolutionary inquiry, which is *prima facie* incompatible with much of what Kuhn said. It will be taken for granted that scientific inquiry cannot be conducted without a background of beliefs about a subject or domain of inquiry. Contrary to popularizations of Baconian inquiry, science cannot be practiced without beliefs that define a domain or a subject of inquiry, constrain and give content to epistemic objectives and specify what constitutes knowledge in that domain, highlight relevant empirical information at hand or to be acquired and provide grounds for ordering problems as to their epistemic or pragmatic importance. Since Kuhn suggested that all of these components change in a scientific revolution, once the old paradigm is challenged nothing is left to guide the development of or the search for elements of the new. Others, such as Laudan (1985) have suggested that in periods of scientific change one of the several components of a paradigm is challenged at a time. Thus the others can be left to guide revolutionary inquiry.

In this chapter we will inquire more deeply into the possibility and nature of revolutionary scientific inquiry, that is inquiry that deliberately seeks novelty in paradigmatic or foundational scientific beliefs. We shall consider more deeply the components of paradigmatic science that might guide revolutionary inquiry and how such guidance actually works. Distinct components of this corpus of scientific and metascientific background belief will be described and the historical contingency of each of these components will be demonstrated. Their bearing upon the research activities of science, that is their normative import, will be of particular interest. Also a general characterization of scientific problems will be given in these terms and conditions for the possibility of deliberate revolutionary research will be formulated. Revolutionary research, that is research based upon the deliberate questioning of paradigmatic components of scientific background beliefs, includes conceptual, ontological, empirical and metascientific inquiries.

2.2. SERENDIPITY AND THE POSSIBILITY OF DELIBERATE INNOVATIVE RESEARCH

Kuhn tells us that extraordinary research results from the gradual dilution of the conceptual and methodological constraints of a failing program in normal science (1970). However, normal scientific research, that is research conducted in accord with a paradigm that consists of an accepted set of values or aims, a worldview, a system of concepts, and a set of preferred experimental procedures, generates the conditions for revolutionary or extraordinary science. Without the accumulation of knowledge of a subject, the division of labor and the focus of normal scientific research, the anomalies that might eventually bring down a paradigm are not likely to be produced. In short, normal science has considerable heuristic value, and its value as heuristics can include the facilitation of revolutionary scientific discoveries as well as expected progress in extending established ideas. However, normal science contributes only to generating conditions for the revolution. It does not contribute intentionally to revolutionary research, whether that be the search for replacement concepts, values, worldviews and experimental methods or whether it be the application of these replacements in a novel program of normal science. According to this conception, the revolutionary discoveries are not planned for or deliberately sought, but generally require changes in the plans and the objectives and criteria used in deliberation. Since Kuhn held that scientific revolutions simultaneously change conceptual systems, metascientific values and experimental techniques, once the constraints of the older paradigm are weakened, there is little to guide the investigator in extraordinary research. On the other hand, the expected products of normal science, according to Kuhn, cannot properly be regarded as discoveries because they lack appropriate novelty, unexpectedness or the element of surprise.

Kuhn's thesis thus implies that there is no deliberate planning or rational action directed toward discovery. Novelty is not one of the aims of normal science. This thesis might be suspected of being a product of verbal stipulation rather than a substantive thesis about science: Anything not produced serendipitiously, that is, anything generated by standard normal research that fits paradigmatic expectations is by stipulation not a discovery. One way of dealing with this suspicion is to produce cases where the novelty of a scientific product is beyond serious debate and yet its production was deliberately planned and sought after. Kepler's efforts at reconciling calculational methods for planetary astronomy with the causal role of the Sun in moving the planets is a planned revolution in astronomy, though his laws of planetary motion are unintended byproducts of his revolutionary activities (Kleiner, 1983). Darwin's search for evidence for transmutation and for genetic concepts that would explain transmutation is another example of a program of contrarian research (Kleiner, 1990). In the light of such cases Kuhn can only reasonably concede that rational extraordinary science is possible.

In sum, Kuhn's position is that novelty in science is generally not generated deliberately or according to plan, but by what has been called *serendipity* (Shapiro, 1986, Kantorovitch and Ne'eman, 1989, p. 513). Serendipity in science consists of discovery by *accident* and by *sagacity* of things one is not in quest of, including (1) an investigator I's seeking to solve problem A and solving problem B instead and (2) I's intending to solve A, but solving B as well as A. Since B is the novel problem and its solution the discovery, it seems to follow that serendipitous discoveries are not products of intent or reasoned plans, where the plans and intent of a research program are embodied in the formulation, choice and investigation of a problem. Serendipity should not be considered a form of 'sleepwalking' (Koestler, 1959) in any literal sense, for sleepwalkers stumble upon things without pursuing any conscious objective. They also need not ever be cognitively aware of what they stumble upon. The investigator who makes a serendipitous discovery is at least pursuing the objectives of problem A and eventually is aware that she has solved problem B.

The solution to problem B seems to come not *causally* independently of the pursuit of A, for in serendipitous discovery it is the pursuit of A that generates the solution to B, though I is not initially aware of this process. This is at least what Kuhn had in mind, though it is not clear from the above definition of "serendipity". Calling the solution to B serendipitous is not necessarily calling it accidental in the sense that the solution just happened and nothing took place to make it more likely. In the purely accidental case the events and processes relevant to B might in *one instance* be occurring in the region of those events and processes relevant to A, and thus I's investigation of A led to encounters with the B-events that then give rise to B. But this seems to be excluded from serendipity as defined above: In serendipitous events the causes of A-events are responsible for making it likely that the B-events come to I's attention, though the encounter with B-events was not initially intended or sought by I. Hence truly serendipitous discoveries are repeatable, whereas truly accidental discoveries are very unlikely to be repeated. A and B might be considered purely coincidentally related, and thus the pursuit of A leading to the solving of B purely accidental, if the *causal* and *intentional* paths leading to A and its solution have nothing to do with the causal and intentional paths leading to B and its solution. The intentional paths leading to a problem are the beliefs and objectives of I that give rise to the problem. The relevant causal paths bringing forth A are those actual circumstances that are causally responsible for the circumstances having occurred that *can* give rise to A in I's mind where only the latter circumstances, as perceived by I, need be relevant to A's presuppositions and objectives. Call these circumstances A-events. The causal circumstances for B, B-events, are similarly defined, but in serendipitous cases I does not anticipate B-events occurring in her pursuit of A. In the serendipitous occurrence of B raising B is probable for investigator I who is attending to A because A-events, unbeknownst to I, make B-events likely. Also it appears that serendipity requires that I raise B when she becomes aware of B-events. Otherwise she would not recognize a solution to B when she encounters it, and hence could not be said to have discovered that solution. It is here that *sagacity* plays a role in B's solution.

Thus serendipity does not necessarily exclude or render inefficacious the deliberate pursuit of epistemic novelty. Serendipitous discoveries occur when the pursuit of A generates unintentionally a solution to B or to A and B. Since B-events are causally linked to A-events, the pursuit of A is a means to the solution of B, though in such cases A cannot be pursued deliberately as such. However, serendipitous discovery of B's solution can be recognized retrospectively as a product of the pursuit of A. Also B itself could be a newly discovered problem or a replacement problem for A and an unanticipated solution to A could be among the B-events as well as a solution to B. The occasion of raising B can be the serendipitous discovery of its solution, in which case B's solution is still not discovered by pursuing B.

Darwin seems to have made serendipitous discoveries regarding specific relations among fossils and specimens brought from South America and the Galapagos Islands. His collecting practices were guided by questions concerning 'centers of creation', possible migratory routes and adaptive varieties that a given species might manifest. The limited adequacy of his collections for his later purposes of establishing transmutation were not an objective or product of deliberate inquiry. However, in the process of examining the specimens he did collect he noticed differences in mockingbirds collected on Charles and Chatham Islands. These differences were the occasion for raising a new question concerning the occurrence of speciation or 'transmutation', and this question motivated him seek to fill in information concerning the particular islands from which other specimens were collected. This episode fits the kind of serendipity described in the previous paragraph. In what follows this episode will be examined closely to determine how it was that Darwin made transitions from intended problems A to revolutionary problems B. The aim of this analysis is in part to evaluate or to qualify the serendipity thesis and, particularly concerning the role of accident and sagacity in this episode.

2.3. CASE STUDY: DARWIN ON THE GALAPAGOS

An attempt will be made here to reconstruct and describe the inquiry that Darwin conducted into the flora and fauna of the Galapagos Islands. The aim of this effort is to clarify reasoning underlying inquiries that produce a shift in the problems addressed and surprising or even revolutionary outcomes. This reasoning bears upon the posing and refining of questions that Darwin addresses to his perceived circumstances and his strategies and steps for finding answers to them. Clarification includes spelling out presuppositions underlying the questions chosen and any inferences drawn, what determines their reasonableness, how such presuppositions might come into question, and how one might deliberately and reasonably search for, discover or choose a replacement for them.

Recent historical research (Sulloway, 1982a,b) has shown that Darwin did not immediately discover geographically representative species, laws of succession and other items of biogeographical evidence for transmutation during his visits to the South American continent and the Galapagos Islands. ('Transmutation' is Darwin's early term for evolution, the thesis that over time organisms change to the degree that new species are eventually produced.) Rather these discoveries emerged only after an inquiry that began perhaps in the last days of his visit to the Galapagos and ended approximately when he began his 'Notebooks on the Transmutation of Species' in July of 1837. This last date is after his conversion to the thesis of transmutation. Furthermore, Darwin's procedures for collection, at least while on the South American mainland and probably until the very last stages of his visit to the Galapagos, were apparently based upon creationist biogeographical background assumptions, which were similar to those of Charles Lyell. Lyell notes the following about species and varieties:

> It is almost necessary, indeed, to suppose that varieties will differ in some cases more decidedly than some species, if we admit that there is a gradual scale of being, and assume that the following laws prevail in the economy of the animate creation: first, that the organization of individuals is capable of being modified to a limited extent by the force of external causes; secondly, that these modifications are, to a certain extent, transmissible to their offspring; thirdly that there are fixed limits beyond which the descendants from common parents can never deviate from a certain type; fourthly, that each species springs from one original stock, and can never be permanently confounded, by intermixing the progeny of any

> other stock; fifthly that each species shall endure for a considerable period of time. (1830–3, p. 166)

Thus, though varieties might occasionally differ more than species, generally there is a barrier that stops variation before species differences are reached. This notion of a barrier suggests that specific differences are either greater or more fundamental to the reproductive processes than are differences between varieties. This suggestion indicates a degree of incoherence in Lyell's species concept, but apparently Darwin did not recognize the conceptual problem this incoherence raises until about a year later. The following remark, written late in the *Beagle* voyage after a visit to Australia, indicates that Darwin had not entirely given up Lyellian creationism at this time:

> It will be very interesting to find from future comparison to what district or centre of creation the organized beings of this archipelago must be attached (1933: 337, entry of Sept. 26–7, 1835)

The procedure of collecting only one pair in each location, a male and a female to provide a record of sexual dimorphism, presupposes that relatively large morphological, biogeographical and behavioral differences separate true species. Also because he believed that varieties and races are caused by adaptations to local circumstances, he did not expect there to be significant variation of Galapagos species among the several ecologically similar islands. Darwin later acknowledges just this point:

> I did not for some time pay sufficient attention to this statement [of the Galapagos vice governor Nicholas O. Lawson regarding Spaniards' recognizing locality of turtles from their distinct forms], and I had already partially mingled together the collections from two of the islands. I never dreamed that islands, about fifty or sixty miles apart, and most of them

> in sight of each other, formed in precisely the same rocks, placed under a quite similar climate, rising to a nearly equal height, would have been differently tenanted.... [B]ut I ought, perhaps, to be thankful that I obtained sufficient materials to establish the most remarkable fact in the distribution of organic beings (1845, 394)

By contrast, Darwin's transmutationist species concept can be found in these passages, probably written in the winter of 1837–8, from the 'Notebooks on Transmutation of Species':

> ...between species from moderately distant countries there is no test but generation (but experience according to each group) whether good species, and hence the importance naturalists attach to geographical ranges of species.
>
> Definition of species: one that remains at large with constant characters, together with other beings of very near structure. —Hence species may be good ones and differ scarcely in any external character. For instance, two wrens, found to haunt two islands —one with one kind of herbage and one with another —might change organization of stomach and hence remain distinct. (DeBeer, 1960, B212f). (Sulloway, 1982b, p. 349)

Darwin's species concept after his conversion to transmutation allows distinct species to be very close in morphological appearances and other characters. Such a conception would require as evidence for distinct species careful observation of habitat, habit and statistics in morphological variation such as average wingspan, and standard deviations therefrom as observed from many specimens. Under such a conception the collection procedures would differ from those Darwin used on the *Beagle* voyage.

Darwin's early taxonomical procedures obscure the morphological and biogeographical fine points needed to argue for transmutation, particularly if the location of the collection is not exactly recorded for each specimen. It is largely by serendipity, not by planning or design in the collection process, that Darwin could obtain sufficient information from his collections in the first part of 1837, after returning to England, to convince himself of transmutation. In regard to Galapagos plants he also acknowledges that due to deficiencies in his taxonomic abilities, much of his collection was 'blind', that is, without specific objectives:

> From my ignorance in botany, I collected more blindly in this department of natural history than in any other; so that certainly it was not *intentionally* that I brought the different species from different islands. If, indeed, I at all noticed their resemblance, I probably collected second and third species as duplicate specimens of the first. (1839: 629)

However, toward the end of his voyage in 1836 he began to suspect that island species might support transmutation. His suspicion was aroused by evidence that certain specimens were either endemic to the archipelago or to particular islands therein. Sulloway dates this passage in the fall of 1836, the last stages of the *Beagle* voyage:

> I have specimens from four of the larger Islands. ... The specimens from Chatham and Albermarle [*sic*] Isd appear to be the same; but the other two are different. In each Isld. each kind is *exclusively* found: habits of all are indistinguishable. When I recollect, the fact that from the form of the body, shape of scales & general size, the Spaniards can at once pronounce, from which Island any Tortoise may have been brought. When I see these Islands in sight of each other, & possessed of but a scanty stock of animals, tenanted by these birds, but slightly differing in structure & filling the same place in Nature, I must suspect

> they are only varieties. The only fact of a similar kind of which I am aware, is the constant asserted differences —between the wolf-like Fox of East & West Falkland Islds. If there is the slightest foundations for these remarks the zoology of Archipelagos —will be well worth examining: for such facts [would *inserted*] undermine the stability of species. (1963, 262)

It was at this time that Darwin undertook an inquiry into evidence that would decide the question (Q1): 'Do species transmute?'. This is a whether-question admitting of two possible answers, 'Yes, they do' and 'No, they don't'. It seems that Darwin viewed the transmutation thesis, the affirmative answer to Q1, as an explanatory hypothesis for certain facts of island and continental biogeography that were emerging in the fall of 1836 and spring of 1837, particularly that certain mockingbird species were confined to particular islands. The explanatory hypothesis can be thought of as addressing one possible answer to a second question (Q2): "Are these specimens distinct species or distinct varieties?" Two possible answers are: 'They are varieties', and 'They are distinct species'. The transmutation hypothesis is an attempt to explain the second of these answers. In the 1830's it was customary to view such local varieties as consequences of powers of adaptation to local circumstances built into every created species, as suggested in the passage from Lyell quoted above. The possible evidential bearing of this second answer on Q1, which contradicted customary belief, is exactly what attracted Darwin's attention to archipelagos in the fall of 1836.

In trying to answer Q2 Darwin had to address a third question, Q3: "Do the variations observable among the mockingbird specimens correspond to variations among acknowledged species?" Q3 is subordinate to Q2 in the sense that answering it is a means to answering Q2 by providing evidence for the specimens being distinct species. Answering Q3 is a means to determining whether or not certain specimens represent distinct species. A bias in Q3 that can be problematic is the kind and degree of correspondence that will warrant classifying the specimens as separate species. Some specific commitment on this matter will be necessary to answering Q3. Another bias arises in assessing Q3 as the means or the best means to answering Q2, an assessment whereby Q3 gains equality in importance to Q2 and Q1. There could be means to assessing species status other than comparing observable differences in morphology, behavior and habitat with like differences between established species. Contemporary molecular biology offers such means. In such cases one could question the import assigned to Q3 and possibly replace Q3 by something else.

In sum, both Q3 and Q2 are subordinate to Q1 because, as Darwin seems to recognize in the fall of 1836, they will provide evidence for whichever answer is offered to Q1. If an affirmative answer to Q1 is particularly important because it undermines the stability of species, and if answers to Q2 and Q3 are means to, perhaps the only means or necessary conditions for answering Q1, then Q2 and Q3 are as worthy of attention as is Q1. Thus Darwin has a strong motive for inquiring into Q2 and Q3. Answers to these questions may be revolutionary in their importance.

Darwin's argument from what he saw to answers to Q2 and Q3, viz. to conclusions regarding whether sets of specimens were distinct species or just varieties, is non-demonstrative and inductive in form. In the case of the birds, the variations in overall morphology, size, habit, habitat, song, beak size, and plumage are consistent with various taxonomic hypotheses, as is indicated by Darwin's own errors and the errors made by more expert ornithologists in England, such as John Gould. As to the mockingbirds, Darwin collected them on only a few Galapagos Islands. With this initial evidence he couldn't rule out the possibility that he had collected only the most extreme forms of one unusually variable species. Three of the four mockingbird forms he collected, those from Chatham, Albemarle, and James islands, are all very similar in appearance. At one time he and even Gould placed them under one species. Darwin never observed these mockingbirds in secondary contact, and hence neither he nor Gould ever had evidence as to whether they would interbreed. Since Darwin's two species of *Rhea* did overlap in their range, he had much more decisive, though perhaps still not completely conclusive, evidence for their taxonomical status, for genuine species never cross.

As to the finches in his collection, Darwin never did understand their biogeography sufficiently to use them as evidence, even as late as in the writing of the *Origin*. Where he did discuss them, he used the transmutation hypotheses in an effort to sort them out. First of all, Darwin, and later Gould, initially identified the cactus finch as "Icterus", a genus in the family of orioles and blackbirds. He also placed a larger beaked finch specimen in the "Gross-beaks". Only a smaller beaked specimen is identified as "Fringilla", a true finch. Also he mistakes the warbler finch for a wren or warbler. Thus he identifies as finches only six of the thirteen on the Galapagos. He also failed to observe any differences in diet among various species and mistakenly concluded that their diets are identical (1841: 99–100). Thus he could never argue that the different beaks were adaptive and produced by natural selection. (Sulloway, 1982a, p. 10)

Once Darwin began to suspect that some of his specimens might be evidence of specific differences among populations on each island and the mainland, he began an inquiry in search of further evidence for this suspicion. Specifically he needed the kind of evidence that would support local species differences and would rule out their being classified as mere varieties.

His procedures are in accord with these steps in diagnostic inductive practice (Wright, 1989): One identifies a question, in this case Darwin's Q2 and Q3, and argues for its importance as an item for immediate inquiry, as Darwin did by presuming them subordinate to an affirmative answer to Q1. One then entertains possible answers to the question and marshalls available evidence for each. Once one has decided which answer is most plausible, the next step is to construct an argument that would rule out the others as implausible. That is, a final step in the procedure is to try to exclude reasonable doubt about the answer chosen.

Darwin recognizes the significance of affirming transmutation, though does not affirm it initially. He simply entertains it as worthy of further research because of its revolutionary significance.

However, its revolutionary significance does lend importance to the search for evidence bearing upon it. The threat to the stability of species highlights the importance of the attempt to determine the taxonomical status of the Galapagos and other specimens, thereby answering Q2, and the importance of finding relevant evidence for this determination, thereby answering Q3. Thus he is motivated to search for evidence bearing upon the taxonomic status of his specimens, evidence concerning their habitat and which ones represent distinct species. For habitat he queried the collections of others, including Captain Fitzroy, a shipmate Harry Fuller and his servant Syms Covington (Sulloway, 1982a, p. 23). For taxonomic status he appealed to the authority of John Gould.

Another strategy of inquiry that contributes to this process of diagnostic inquiry can be described as an application of the traditional 'method of analysis' (Hintikka and Remes, 1974). In the search for evidence one can 'reason backwards' from the conclusion by iterating these questions: If *h* is one alternative answer to the favored question Q, what is evidence *e* that would support *h*? This question can be divided into two subordinate questions: Q4: What kind of evidence would support *h* if it were available? Q5: Once we have determined that *e* is appropriate for supporting *h*, how do we make *e* available to our cognition, and, Q6, does *e* hold in the circumstances under consideration? The first of these subordinate questions, Q4, asks for evidential relations whereby possible *e* can support *h*. The last, Q6, asks whether *e* is the case and the second, Q5, how one might go about investigating, e.g. by experiment or other forms of investigation, such as museum or library research, field expeditions, or consulting experts, to determine whether e is the case. Q5 is subordinate to Q6 in the sense that Q6 is the objective, to assess the truth of *e*, and answers to Q5 are judged in terms of their efficacy as means to answers to Q6. These last assessments are often, though not always, made in the light of answers to Q4, the specification of evidential relations. Thus if comparing the differences among his specimens to differences among established species is a means to determining the taxonomic status of his species, to decide the latter question (Q2) Darwin must gain access to more detailed knowledge of established species. Under the circumstances at the time consulting authorities such as Gould and Owen would be a reasonable means to such knowledge.

It should be clear from the above examples that the transmutation hypothesis had significant implications regarding Q4, Q5 and Q6. Darwin's research project in 1836 and 1837 included the determination of these implications and the use of them as guidelines for further research. If his specimens might be genuine species, then much more specific information about the geographical extent of their habitat is required, for true species may be much more localized in their range than was previously thought. Also, to determine the taxonomic status of his specimens more specimens would be needed, something that he realized probably in response to expert information provided by Gould and other taxonomists in England, where this information led to revisions in Darwin's initial answers to Q4. Additional specimens can contribute to determining ranges of variation within populations, where distinct ranges of variation can provide evidence of specific differences. Gould was probably the source of the wren example appearing in Darwin's transmutationist species note. Gould's authority can illuminate specific requirements on ranges of variation, an answer to Q3. Gould's access to museum collections, which might contain further specimens from the same localities as or neighboring localities to the origins of Darwin's own specimens, can be exploited as a means to answering Q6 once Gould has provided a satisfactory answer to Q3. Other strategies for answering Q5 and ultimately Q6 include undertaking further expeditions to the relevant localities and collecting larger samples from the populations there. Such expeditions were in fact undertaken beginning with Habel's in 1868 and followed by eight more before the turn of the century. These expeditions illustrated a shift in the collection procedures due to revised answers to Q4: Whereas Darwin collected 31 finches and a total of 61 birds from the Galapagos, Habel collected 460, Georg Bauer 1,100 in 1891, Charles Harris 3,075 in 1897 and the California Academy of Sciences 8,691 in 1905–6 (Sulloway, 1982a, p. 40).

According to this model of inquiry as based on diagnostic induction, the shift in evidential relations or answers to Q4 need not be considered biased in favor of transmutation, question-begging, or even bootstrapping. The method of analysis only asks for the evidence required for a proof of a conclusion, and in so asking it attends to the conclusion without assuming its truth. Hence, though a hypothesis must be conceived or thought of to discover relevant evidence for it, it need not be believed for that evidence to be conceived or sought.

Darwin's initial investigations of Galapagos species were guided by Lyellian creationist biogeography. Among the background assumptions guiding his choice of questions to pose there in the field is his assumption that species are introduced at a 'centre of creation' from which they migrate to ecological limits. These are limits beyond which, due to adverse geography or climate, competitors, lack of food, disease, plants and animals cannot spread. Ecological background assumptions include a number of causal relations that may be appealed to in plausibility judgments made in elaborating answers to questions under investigation. These plausibility judgments may include estimates of prior probabilities, that is, probabilities of events in the absence of relevant empirical evidence. One such estimate is of probabilities that mice or tortoises were brought to an island by humans in contrast to probabilities of their migration there. The former can be considered probable and the latter improbable on grounds of historical, ecological and biogeographical background beliefs without reflection on actual evidence, e.g. that humans actually visited the island in question with the tortoises and mice on board. These background beliefs may also provide evidential relations needed for estimating posterior or conditional probabilities, the probability that given e, h will occur. For example, a measure of the width of the channel can be the evidential basis, e, of estimating the relative probability of a bird's crossing, where evidential relations include reflections upon the birds flight capabilities, endurance, prevailing winds, etc. Lyell also views varieties as natural adaptations to varying conditions that can be expected to occur in neighboring regions as consequence of colonization and adaptation. However, according to this conception, one might also expect there to be different ecological circumstances in which varieties do occur. Such expectations provide likelihood relations on which posterior probabilities can be calculated using Bayes' theorem. On the basis of such relations Darwin did not expect to find significant variation in species among the Galapagos Islands. Also, given Lyell's assumption that distinct species must be introduced by distinct acts of creation, and acts of creation occur to fill empty niches left by extinction, the accessibility of the various Galapagos Islands to their bird populations led Darwin not to expect to find different but related species, species of a common genus, on the various islands of the archipelago. These expectations influenced the questions he posed in a negative way: He simply initially did not ask 'What are the local varieties/species on each island?'

One plausible source of Darwin's suspicion that neighboring variants are true species are two forms of *Rhea* he observed earlier in Patagonia. Apparently he did not think of applying this clear cut example of locally representative species until he reexamined specimen mockingbirds in collected in the Galapagos in the fall of 1836. These then brought to mind foxes in Falklands and other examples that suggest that the line between variety and species cannot be clearly drawn. At this point Darwin undertakes Q1, 'Has transmutation taken place?' Thus he begins his investigation into the possibility that many specimens from neighboring habitats are genuine species. In carrying out this investigation he must examine the current species concept to determine evidential relations bearing upon possible answers to Q1, that is he addresses Q3 as a means to determining the taxonomic status of his specimens, Q2.

Thus Darwin has good reason to search more deeply into the *Beagle's* Galapagos collections. His search includes an appeal to the authority of John Gould, Richard Owen and others, from whom he is able to draw this conclusion:

> I may observe, that as some naturalists may be inclined to attribute these differences [among the three island forms of mockingbird] to local varieties: that if birds so different as

> *O. trifasciatus* and *O. parvulus*, can be considered as varieties of one species, then the experience of all the best ornithologists must be given up, and whole genera must be blended into one species (1841: 63–64)

This is Darwin's final and seemingly decisive objection to the rival answer to Q2, that the various specimens are only locally representative varieties, not species. By June of 1837, Darwin has found so many locally distinct but related species, including related fossils, that evidence for transmutation seem to him overwhelming.

Darwin's strategies of inquiry may be summarized as follows:

1. Questions are highlighted as to their scientific importance on grounds of the possible impact of some of their answers on beliefs constitutive of a research tradition. The stability of species is one such belief. The importance of such questions also depends upon the importance of the belief: If giving up such a belief has wide ranging implications, e.g. it requires changing a large number of those in the network of paradigmatic beliefs, then answers to questions that challenge it are even more important.

Feyerabend's contrarian strategy (1974) recommends that such beliefs be sought out and that one attach high priority to their criticism. However, we cannot expect science to advance if all researchers undertook such a strategy all the time. As a universal rule, Feyerabend's contrarianism has no heuristic value. On the other hand, if there occurs *prima facie* evidence that a highly significant belief constitutive of a research tradition might be false, formulating and investigating alternatives to it would seem highly promising in regard to producing new knowledge. For heuristic value Feyerabend's contrarianism must apply only where there is good reason or well founded suspicion of traditional beliefs and must not be applied arbitrarily. Initial suspicion that there are geographically representative species highlights the importance of Q1 and all of its subordinates once the revolutionary importance of that concept is perceived.

2. A second strategy is to divide questions into subproblems,

where these include problems of finding means to answering a preferred question. Means that appear necessary to answering a question, as might but cannot always be determined by a logical analysis of the question, take on the same importance as the original question. Optimal means, or those which seem *prima facie* most likely to produce answers are valued according to the value attached to the original question and according to their promise in producing answers. Those that are less promising will be less valued among those likely to solve the problem. The import of Q2, determining the taxonomic status of the various specimens is derivable from Q1 by logical analysis of one of its answers, the doctrine of transmutationism. Transmutation can occur just if speciation has occurred. Thus the specimens are an instance of transmutation just if they are specifically distinct.

Q3 is based upon a more substantive view regarding what distinguishes species: The barrier to crossing correlates with observable morphological, behavioral and ecological features of a population. In cases where sympatry does not occur one must rely upon this correlation. Hence by comparing the specimens' differences in morphology, habit and habitat to the differences in recognized species, one may answer Q2. There could be other means of determining from observable characteristics the taxonomic status of the specimens. Hence the import of Q3 is based upon a presumption of the above correlation, which could be false.

3. One heuristic for dividing questions is based on method of analysis which generally prescribes search for evidence for hypotheses entertained. It is by this method that Q3, 'Are the specimens comparable in their differences to recognized species?' is derived from consideration of a possible answer to Q2, the taxonomic status of the specimens is that they are distinct species. Analysis also divides Q3 into two steps, Q4 the determination of what kind of evidence is relevant, Q5, how to produce cognitive access to the evidence, and Q6 the determination of its truth value. This is reasoning backwards: If transmutation were the case, how would I go about finding evidence in support of it? What would constitute evidence? How can I produce or gain access to that evidence? Is

that evidence as expected from the transmutation thesis?

4. Practical considerations in the search for evidence further divides Q3 giving Q5, which asks for means of implementing search for evidence. Answers to Q4, the evidential relations and the kind of evidence sought, constrain but do not determine answers to practical questions of producing, generating or otherwise gaining access to the evidence sought.

5. Darwin follows out implications of transmutation thesis regarding what would constitute evidence for it. Here he is applying the analytic method. This procedure leads him to look for specific differences in smaller variations of habit, morphology, etc. Thus he must search for more specimens and improve the accuracy of range determinations. Gould's, Richard Owen's and other experts' descriptions of specimens and fossils provides so many cases of representative species that Darwin finally accepts transmutation, and then undertakes a search for its explanation, which occupies the next 18 months.

6. A number of the steps Darwin took are in accord with Nickles' conditions on the rationality of constraint violation (1980), though not all of his conditions are relevant to this case. Received constraints based on Lyell's biogeography are revised in light of requirements for the transmutation hypothesis, facts about specimens from *Beagle* voyage, and the practice of expert taxonomists in determining taxonomic status. Such constraints are fallible, and should always be regarded as such. Darwin's *prima facie* evidence for transmutation made it reasonable to question Lyell's constraints on species variation, the subject of Q1. Darwin was initially ignorant of then current taxonomical practice but discovers a lot about it in search for evidence for transmutation. When norms of taxonomical practice are applied in context of Lyell's biogeography, particularly ecological theses governing migration and range, evidence emerges from *Beagle* specimens for representative species, and in turn for transmutation.

It should be clear that this episode is not just a factual discovery that is described using received concepts. It also requires the revision of these concepts in the search for evidence. It is also not an episode in which new concepts are invented. They were already available in the practice of taxonomists such as Gould and in Lyell's biogeography. However, the conceptual novelty of this episode was the incorporation of these established taxonomical concepts into the practice of biogeography and the removal of the 'barrier' between species and true-breeding races. This conceptual shift removed the incoherence in Lyell's description of a species concept quoted above.

2.4. A GENERAL TAXONOMY OF SCIENTIFIC PROBLEMS

In this section several components of the cognitive background to scientific research will be distinguished and described. Various components of this background are presupposed by various aspects of scientific research, some of which would be described by Kuhn as 'normal' but others would have to count as 'revolutionary'. This analysis is a variant on numerous others (Toulmin, 1967, Kuhn, 1962, 1970, Maxwell, N. 1974, Laudan, 1977, 1985, Kitcher, 1985, 1990). The argument for this analysis will be in part that it reflects various developments in the history of science. Also it will be shown that it underlies a plausible though broadly general taxonomy of scientific research problems.

1. Background Ontology B: Scientists generally operate with background beliefs extending beyond their own field of interest. They hold beliefs about what there is in the Universe beyond their field of study and beliefs about how these other entities, circumstances, events and processes are structurally and causally related to what it is they study. Such beliefs play an important role in scientific research, particularly but not exclusively in the search for new ways of observing what goes on in a given domain of scientific interest. Insofar as these beliefs go beyond their own field, the background beliefs of a given investigator often, but not always, contains little in detailed models for various components of the Universe, though it does contain general information about what is there, and what characteristics distinguish the various components of the Universe and how they relate to the observable and unobservable parts of the locality occupied by the investigator. The extent of these beliefs is various among different individuals in a given discipline, and normative expectations peculiar to each discipline regarding the content and extent of this background varies from discipline to discipline and even varies within disciplines. Sometimes a field will significantly advance when somebody enters with new or exceptionally detailed knowledge of a related field. Such happened when crystallographers entered biochemistry in the early stages of the development of molecular genetics (Judson, 1979).

Such background belief systems have variously been called 'world views' and 'metaphysical presuppositions'. The latter terminology may not be entirely felicitous because it suggests something entirely apart from science, perhaps infallible and not responsible to the same epistemic authority or metascientific criteria that science is responsible to. 'World view' or *Weltanschauung* might be an acceptable choice of terminology if it is remembered that these background belief systems include beliefs about the entire universe as well as more particular beliefs about various components of the universe. 'Cosmology' is also an acceptable term if it is remembered that B usually contains what would be in some respects both more and less than the belief systems of practicing cosmologists. Practicing cosmologists will have deep knowledge of physics as it occurs under cosmic conditions but often have little detailed knowledge of conditions prevailing on a planet, particularly ecological conditions on the Earth, for example. On the other hand ecologists know enough cosmology to understand in some detail the causes of diurnal and annual variation in sunlight, conditions in the atmosphere under which greater ultraviolet penetration of the atmosphere might occur, etc. They may also have a broadly general knowledge of physical and geological conditions on other planets, which could serve as a source of conjectures regarding the physical conditions for terrestrial ecological stability, and an even more broadly general knowledge of the rest of the Universe, perhaps only as a means of locating their domain of interest, the terrestrial sphere. In the following this component of scientific knowledge will be referred to variously by 'B', 'background cosmology', 'ontology' and *weltanschauung*, as needed for stylistic variation.

Perhaps the closest we can get to defining B's content is to think of it as a general cosmology having a rough but not exhaustive consensus among scientists and scientifically aware lay persons. As the above examples indicate, it also should be allowed that different individuals will have more developed beliefs in various different aspects of this ontology. This variation can be a heuristic advantage when a community does not know in advance which such relations might be productive of fruitful new insights. B will contain general descriptive knowledge of the universe and its contents and history wherein certain domains of inquiry can be localized and related to others. Thus we might localize terrestrial geology or the Earth's biosphere within the Solar System, etc. The descriptions are in terms of a general conceptualization that combine educated commonsense with some and various deeper knowledge depending upon expertise. As new conceptualizations are introduced into scientific disciplines and become common knowledge some of these descriptions will be revised essentially and deeply. A biologist doing biomechanical studies of quadruped locomotion may possess a general commonsense descriptive knowledge of the cosmos with a rather detailed knowledge of classical mechanics, particularly statics, but might also possess a detailed knowledge of wave optics which gives her a unique advantage in developing novel cinematographic methods for biomechanical observation. Other components of B include, e.g., beliefs about the number of Solar planets and some of their distinguishing characteristics, the accessibility of their motions to mechanical description and explanation and to terrestrial observation, the current geographical arrangement of the continents, deserts, mountain ranges and oceans with respect to various climatic zones, rough or detailed knowledge of recent theories of continental movement, organizational and compositional hierarchies of living organisms, and some conception of descriptive and explanatory procedures appropriate to each level, etc., etc.

B thus contains ontological components, e.g. current general conceptions of entities, causal process and ontological hierarchies of structured assembly for atoms, chemical elements, compounds, macromolecules, biochemical processes, tissues, organs, organ systems, groups and structured populations of organisms, of time and space in cosmic and local applications, as well as atomistic, plenistic and organismic patterns of organization. B is not 'metaphysical' in the sense of being beyond epistemic reproach. There are numerous examples in the history of science in which more or less radical, though never total, revisions have occurred in this background *weltanschauung*. Ancient cosmology formulated by Aristotle is bound by a relatively small sphere, divided into celestial and terrestrial domains in which processes are governed by different kinds of law as well as being descriptively distinct. From the Moon outward, the heavens are eternal, move in circles and never undergo any qualitative, quantitative or relational change. Terrestrial objects generate and corrupt in eternal reproductive cycles, under growth, changes in outward form, quality, etc. In spite of cosmological and conceptual differences which have been described at length by historians and philosophers of science (e.g. Kuhn, 1959, Feyerabend, 1962, 1974), Aristotelians share with other figures such as Ptolemy, Tycho, Kepler, Newton, and even the contemporary historian or professor of astronomy, several sophisticated descriptive concepts for what can be observed in the heavens with the naked eye, e.g. the apparent celestial sphere, the ecliptic, celestial poles and equator, variation of appearance with terrestrial movement, the rectilinear projection of visual images, descriptive and ostensive identifications of the sun, Moon and five 'planets'. *Appearances*, like mirror images and mirages, are as much part of the Universe as that which appears, though there may be significant differences across history in what one takes an appearance to represent. Aristotle's realistic conception of the celestial sphere and geocentric conception of planets are significantly different from other concepts emerging in the history of astronomy, especially from the 16th century on. Newtonian planets are mechanical bodies like the Earth or projectiles on the Earth, and in Newton's case the extension of 'planet' does not include the Sun and Moon, as it does for the geocentrist. Galileo and Kepler considered the Sun a unique body in a finite spherical cosmos, but Descartes and Newton considered it one of many stars in an unbounded universe.

To further illustrate deep differences in B, the Newtonian cosmos is infinite in extent and in worlds (material contents), undifferentiated between celestial and terrestrial in regard to the application of fundamental descriptive concepts. These concepts are the mechanical 'primary qualities' some of which, particularly 'force' and 'mass' are implicitly defined by Newton's laws of motion, which govern all processes, microscopic and macroscopic. (See Newton's 'Rules for Reasoning in Philosophy' in Book III of the *Principia*, particularly Rule III, Thayer and Randall [1953] pp. 3f, and his 'Queries' in the 4th edition of the *Optics*, especially Queries 30 and 31, Thayer and Randall [1953] pp. 158–179, for these views and their methodological ramifications). Newton views the domain of his mechanics as universal, including terrestrial and celestial, microscopic and macroscopic bodies. More recent cosmologies are differentiated into hierarchically organized descriptive domains related by composition and organization, e.g. sub-atomic, atomic, molecular, cellular, etc. In contrast to Newton's view, these domains are now described with significantly different sets of descriptive concepts, including the quasi-mechanical complementary properties 'position', 'momentum', and 'spin', and non mechanical properties such as 'strangeness', 'color' and 'isospin' for fundamental particles. For higher domains Newtonian concepts represent only aspects or partial descriptions of objects with other scientifically important attributes, quite to the contrary of the traditional doctrine of primary qualities. Thus the domain of mechanics is far more limited than was thought by many in the 17th and 18th centuries.

To summarize, B has many functions in scientific inquiry, among which are to provide the knowledge necessary for distinguishing separate domains of inquiry, to provide a background of interdomain relations that might be exploited (i) for explanations otherwise unobtainable (Maull, 1977, Darden and Maull, 1977) and (ii) for possible evidential relations to domains that can be manipulated, observed or both. B also provides a background of anticipated uniformity, and specific causal and structural relations that licence and delimit inductive and other inferences: In Aristotle's cosmos the kind of inductive inference that Newton used, from the observable laws governing terrestrial bodies such as projectiles to unobservable causes of observable orbital motions in celestial bodies, such as gravitational forces acting on the moon, Mars and Jupiter's satellites, would not have been permitted. Aristotle's 'Lunar Dichotomy' divides the cosmos into two domains in which different laws governing the locomotions and other changes of substances hold. One simply cannot infer that like causes move projectiles on the earth and the moon in its orbit in Aristotle's cosmology. Similarly, Newton's inductions using Rule III from primary qualities of macroscopic to those of microscopic bodies and atoms would not be considered valid in the context of today's hierarchical conception of compositional organization and emergent properties of matter. Thus B plays an important role in sanctioning inductive and analogical arguments and heuristics in scientific problem solving, a role that will be further discussed in Chapter 6 below.

B can also licence deductive inferences: As pointed out earlier the 'discoverability' of newtonian gravity depends upon the law of inertia. This law is appealed to in Newton's deductions from Kepler's laws that central forces constrain the moon, Jupiter's satellites and the various planets and that these forces are inverse square forces.

How it is that the explicit or implicit domain structure in B can be revised without begging questions regarding inductive and other scientific procedures is an old and important methodological question. In the case of terrestrial→celestial inferences, Aristotle's lunar dichotomy, which blocked such inferences, was undermined by evidence for the Copernican system and collapsed as a consequence of Galileo's telescopic discoveries at the beginning of the 17th century. First, the dichotomy was never applied to light: The Pythagoreans, Plato, Aristotle and others responsible for formulating and defending the beliefs underlying the ancient 'two sphere' universe (Kuhn, 1959) all implicitly assumed that terrestrial and celestial optics are the same in their arguments for the central location of the Earth, its sphericity, etc. Secondly, it is incompatible with the Copernican doctrine that the Earth is a planet, and insofar as this doctrine is supported by Copernicus' arguments from coherence, celestial-terrestrial uniformity for dynamic laws is not assumed. Copernicus did, however, assume celestial-terrestrial uniformity for dynamic laws in his doctrines of circular inertia, which apply to terrestrial bodies, the earth and the planets. This assumption is perhaps only weakly supported within the limits of Copernicus' own perspective by presumed universality of optics, though as an essential component of his defense of the planetary Earth against contrary empirical evidence, such as the 'tower experiment', it does seem to beg the question of the truth of Aristotle's celestial dichotomy. Galileo's telescopic observations of lunar mountains, sunspots and Jovian satellites further undermine the lunar dichotomy without begging the question about the universality of dynamic laws, because the observations depend upon universal optics, which was never at issue. As such they weaken Aristotelian grounds for this dichotomy. His observations of Venus' phases proves that Venus is a solar planet. This proof is a *deduction* based upon the assumed universality of optics. However, in the light of Tycho's geocentric model, this observation does not prove heliocentricism. However, the undermining of Aristotle's lunar dichotomy just described, particularly that resulting from Galileo's discoveries, which are also proofs based on optical principles, of Lunar geology and Solar meteorology, helped to remove geocentricist objections to heliocentric cosmology and rendered universal dynamics a possibility without proving or providing decisive evidence for either one. Specifically, these discoveries failed to render Tycho's system any less believable, possibly by making it more *ad hoc* than it would have been otherwise: Since the moon circles the Earth and has a geology like the Earth, why should the Sun, which circles the Earth, not have, again like the Earth, a meteorology?

The methodological upshot of this history is this: By 1610 Aristotle was proven decisively wrong in his dichotomizing dynamics into celestial and terrestrial forms. Thus it was made plausible without circularity that dynamic uniformity exists between the heavens and the Earth. Though heliocentricism might have lent support to a universal dynamics it did not as of 1610 because there was no decisive evidence in its favor. However, not all ancient and medieval thinkers followed Aristotle in this dichotomy: John of Philopponous around 600 A.D. defended the *impetus* theory of projectile motion and Oresme in the 14th century applied it to heavenly motion but retained geocentricism. Hence universal dynamics (excluding gravity) could be made compatible with both heliocentric and geocentric cosmologies, and the destruction of the lunar dichotomy made it reasonable, without begging unsettled questions, to apply the same laws governing the causes of motion (gravity excluded) to projectiles and planets (in either the heliocentric or the geocentric sense of the latter term). Thus one has licence to infer inductively from terrestrial experiments establishing dynamic causes of certain effects to unobservable celestial causes of similar effects.

2. Epistemic Criteria M. Epistemic criteria are metascientific and sometimes epistemological beliefs concerning the characteristics an observational or theoretical proposition should have to be worthy of acceptance or belief, or worthy of entertaining for further development as a candidate for belief.

Epistemic criteria have been taken by some to specify *truth conditions*, the actual conditions that prevail or are probable if a given proposition is the be accepted as true. Truth conditions can be viewed as implications or consequences: What is the world probably like if p is true? These may include but need not be confined to *empirical content:* What would we expect the observable features of the world to probably be like if p were true? Few if any truth conditions so defined can be known apart from a set of ontological beliefs B, for it is largely in the light of these beliefs about the various parts and aspects of the Universe that the world-wide implications of some p can be discovered. Thus we should expect that the truth conditions for some p, such as the proposal that the Earth moves, would be different in an Aristotelian cosmology from what they are in a Newtonian cosmology. The motion of the Earth in an Aristotelian universe would cause 'gravity' to be different from what is observed, and would produce catastrophic earthquakes, winds and dislocations of detached bodies that have never been observed. (N.B. These happen to be the empirical content of the possible motion of the Earth in an Aristotelian universe, but, as we have claimed, truth conditions need not be limited to observable consequences. The truth conditions of the Earth's motion in a Newtonian universe are not all observable, for though its motions relative to other bodies are observable, its state of absolute motion is not.)

Because of this relativity of truth conditions to substantive beliefs in the background *weltanschauung*, metascience is best taken as containing prescriptions for truth conditions, for example: If *p* is to be accepted as true, then we must also accept as true or as probably true its truth conditions as appear in the light of our background ontology B. Positivism contained the metascientific requirement that truth conditions be empirically accessible, but that requirement is too strong because empirical accessibility depends upon current knowledge of evidential relations and technologies bearing upon their implementation. A more defensible metascientific principle demands that *p* and its truth conditions should form a coherent non *ad hoc* whole, that is, *p* and its truth conditions should be brought together within a network of internal and intelligible relations. In the light of the above remarks this prescription amounts to the demand that background ontologies be coherent. It demands that we either make our cosmology as coherent as possible or choose the most coherent among available cosmologies. The demand for coherence encompasses observable as well as unobservable propositions, and insofar as it includes the former propositions it is also a metascientific demand that background ontologies be empirically adequate. As such they cannot be called 'metaphysics' in the traditional sense.

Other metascientific attributes include desired characteristics of concepts or propositions believed relevant to their truth, and these typically include in some sense simplicity, explanatory power, generality in scope, internal and external coherence, empirical content and specificity and precision (Popper, 1958, Maxwell, 1974, Laudan, 1977, Thagard, 1978). Much philosophical discussion is devoted to elaborating the content of these desiderata and to their justification. Some have held that they are conventional or justified by arbitrary convention, perhaps a conventional stipulation that demarcates science from other intellectual endeavors (Popper, 1958). Kuhn's historicism might be considered conventionalist though pluralistic. Each paradigm has its own metascientific elements and the decision to embrace a paradigm is like a holistic conversion experience or a gestalt shift. Once and only if one steps into the circle one sees it all in a coherent whole. Others attempt to derive some desiderata from others, e.g. simplicity from empirical content (Popper, 1958, Rosenkrantz, 1977).

A pervasive problem regarding metascientific criteria is linking them to a plausible conception of truth for science. On the one hand, if one assumes the correspondence concept of truth, then for coherence, explanatory power or simplicity to be truth criteria, we must assume the world or the domain of inquiry to have these properties. However, we cannot establish this assumption without accepting as true certain simple, coherent and explanatory theories. We are caught in a vicious circle if we must appeal to simplicity, etc. in appraising as true these theories. One way out of the circle is to give up correspondence as an account of truth. If we put in its place coherence, then the justification of these criteria in the case of coherence is trivial and in the case of simplicity or explanatory power is a matter of reducing them to coherence. However fictions are also coherent and can provide simple explanations. Accordingly this move fails to satisfy those who believe that science aims at stories and explanations at least some of whose terms refer to actual events, entities or processes. A third option is to view the aim of science as pragmatic, the achievement of beliefs that foster actions that conduce to achievement of our goals. In this case, it is clear that empirical content has pragmatic import because observable consequences are those which can influence us causally. These same consequences thus are also likely, if anything is, to be beneficial or harmful. However, it is not immediately clear how explanatory power, simplicity and coherence have pragmatic virtues. However we have seen Peirce's suggestion that evolution selects those with instincts, possibly with innate criteria, for picking the correct hypotheses. Among these selected criteria might be certain of the above mentioned metascientific desiderata.

Laudan (1984) has suggested that metascientific values can be established by being exemplified in the history of science. If a simple theory with great explanatory power and empirical content appears in history this shows that the objective of producing theories with these attributes is not *utopian* or unachievable. History might also tell us that simple theories with high explanatory power are very often empirically successful, and hence the probability that a simple theory is true can be based on historical frequencies. Similar cases might be made for the imposition of other metascientific requirements: The success of Newtonian mechanics whose principles are invariant under changes of velocity as computed by 'Galilean transformations' warrant the extension of such invariance to other theories, such as Maxwell's electrodynamics. When Maxwell's theory turned out not to be invariant under Galilean transformations, Galilean relativity was given up as a regulative principle. However, the success of Maxwell's electrodynamics whose laws are invariant under 'Lorentz transformations' warrants the extension or these transformations to classical particle dynamics and to quantum mechanics.

It is thus that we might establish certain methodological values as instrumental to others which are directly sought in science. One seeks simple, explanatory or invariant theories as a means to obtaining theories that later turn out well confirmed (Salmon, 1966). The science of astronomy was founded in the ancient world in response to the practical need of time keeping, that is having a method of forecasting how many days or months it will be before the spring equinox occurs and having means of determining the fortunes of individuals or the state. Greek astronomers of the sixth and fifth centuries B.C. discovered that forecasts of apparent positions of the Sun and stars were facilitated by taking as a reference frame a sphere concentric with the earth on which coordinates could be defined and used for location and plotting motions. Setting this concentric sphere in motion also tied together an indefinite number of apparent diurnal motions of the stars in a manner that made the sizes and locations of their orbits readily understandable. By reference to this sphere one could compute the ascension of the North Star at various latitudes, and thus partially fill the need for determining one's location at sea. Thus practically useful devices also have explanatory power, and the converse relation might be expected in other instances. Furthermore, Copernicus' simpler arrangement of the planetary system, simpler in the sense that it did not contain five epicyclic motions that coincidentally paralleled the Sun's motions, proved to have greater explanatory power, in explaining the different appearances of inner and outer planets. It also had greater empirical utility in allowing the computation of the relative distances of the planets from the Sun. Finally, Galileo showed that some of its predictions could be confirmed with the telescope and Newton's third law of motion implied that the Sun's motion must be vastly smaller than that of any of the planets, including the Earth. Thus a simpler theory proved to be superior to its predecessors in explanatory power, computational utility, and was empirically confirmed and shown coherent with inductively established mechanical laws. The Copernican revolution thus provides a paradigm vindicating the valuation of theories that eliminate causally unconnected coincidences in the movement of bodies, whether planets or other.

Once metascientific desiderata are established by paradigmatic exemplar, they can serve as objectives to be sought in further scientific investigations. Darwin explicitly claimed (DeBeer, 1960, B110f) that he sought for natural history or biogeography 'simple' and 'sublime' laws like those governing planets' motions. He compared special creations to God's commanding each position in each planetary orbit. Accordingly, one law which when applied in various ecological circumstances would predict various evolutionary changes can be compared to laws of planetary motion which, when applied to the peculiar circumstances of each planet, predicts the correct motion of that planet.

If the current state of knowledge falls short of some proffered metascientific objective, than a problem arises whose goal is to modify that knowledge state so that the objective is achieved. Accordingly, one line of argument in support of such a problem is to show, as Laudan suggests, that such objectives are attainable and to show that they render probable other desired objectives, such as the attainment of beliefs that are true. Darwin thus could argue for his problem of obtaining a non *ad hoc* law governing species on the grounds that similar goals have been achieved by astronomers and that these achievements produced broadly based confirmations of astronomical theory, that is they resulted in substantial arguments for their truth. Given that the natural historian has similar epistemic objectives, she should also seek 'simple and sublime' laws for natural history.

M might also be thought of as consisting of a number of metascientific values to be sought as opposed to a number of rules to be followed in scientific research (Kuhn, 1977, Chapter 13). A plurality of values can be partially achieved by different research programs so that on one count the first program is better than the second and on another count the second is better than the first. In such cases the choice of research programs is underdetermined by the current state of achievement of metascientific objectives.

Consider rules corresponding to these values: 'One should accept only theories with maximum empirical content', 'One should accept only theories with maximum simplicity and minimum *ad hoc*ness', 'One should accept only theories with the greatest confirmation', and so on (Cf. Feyerabend, 1974). Thus stated the rules are mutually incompatible unless all necessary conditions are met simultaneously. Accordingly, if we have a simple theory that is not well-confirmed, the latter rule will have to be suspended if the theory is to be accepted.

Values or metascientific desiderata are plausibly considered additive in their epistemic import: A simple well confirmed theory is better than one that is just well confirmed, etc. Thus considered, one can argue that seeking a simpler formulation of a well confirmed theory or confirmation of simple theories are reasonable problems or research objectives (Kleiner, 1983). Thus one might reasonably pursue a qualified contrarian strategy, as suggested earlier. If a contrarian theory or a research program fulfills one epistemic objective then it is heuristically reasonable to pursue it further, in particular to seek what it takes to make it fulfill other epistemic objectives. This qualification renders Feyerabend's 'anything goes' heuristically tractable.

3. Conceptual systems C. The formulation of background beliefs B, and the background beliefs themselves, make use of and contain one or several systems of descriptive concepts C, C′,.... These include the kinematic and dynamical concepts of Newton, which in his view served as an ideal scientific language for all of B. More pluralistic and anti-reductionistic worldviews generally will admit a plurality of conceptual systems applicable within the various domains and hierarchies postulated in B. Thus at the level of the biosphere that is divided into reproductively isolated populations, species concepts are applied to those reproductive groups that cannot exchange genes with one another. Species concepts have associated with them clusters of 'default' expectations about morphology, habit, habitat, sexual and juvenile dimorphism, as has been illustrated in the discussion of Darwin and his colleagues above. Components of mechanical systems and biological species alike are defined by clusters of determinable attributes which in turn are characterized by conditions and constraints on their possible values.

Mechanical 'bodies' and biospecies are natural kinds that appear in distinct domains of a hierarchy, and they carry with them the normative expectation that other members of the respective domains should have like characteristics: In the case of mechanical bodies it is appropriate to ask 'What is its location with respect to a given frame of reference?' '...its velocity...?' '...its mass...?' etc., where these questions will be different in their constraints respectively in classical and relativistic conceptual systems (Kleiner, 1970). In the case of species, it is appropriate for a new member to ask for morphology, habit, habitat, etc. with the kinds of constraints discussed earlier.

Thus the system of domains and natural kinds in B allocates distinct conceptual systems to its constituent domains and natural kinds and these concepts give content to descriptive questions arising concerning membership of each natural kind. In general, when one asks for the identity of an individual object or organism one is asking (i) for its location in the natural kind hierarchy and (ii) for the determination of determinable characteristics attributable to the individual once its kind is known.

Changes in worldview can have as a consequence changes in concept, and conversely. Changes in worldview can include changes in the extent of the domain of expected applicability and changes in the range of values taken by a concept. Darwin's merging varieties and species required significant changes in expected habitat, ranges of morphological variation and habit, as illustrated above. Newtonian mechanical concepts are now restricted to ponderable bodies with actions (energy $\times$ time) significantly larger than Planck's constant (6.62×10^{-27} erg. sec.) and relatively low velocities ($v/c << 1$, where c is the velocity of light, v the velocity of the body in question) in relatively weak gravitational fields. Changes in concept with consequent changes in worldview include the following: Daltonian concepts of chemical element and compound require that the domain of 'atoms' and 'molecules' be distinguished as bearers of chemical properties, viz. powers of forming compounds in specific atomic or volumetric ratios with other compounds. Earlier 'atomisms' offered no possibility of identifying quantitative descriptions for their atoms. Biological properties of form, reproduction, nutrition and accompanying behaviors defines a subject domain of living organisms for Aristotle and later naturalists. In the 17th century these properties are extended to infusoria and in the 19th century to cells. Thus the biological worldview acquired microscopic domains. The biospecies concept requires that populations of interbreeding organisms be considered separate components of the biological universe for purposes of studying biogeographical distribution and evolutionary history.

4. Laws L associated with concepts. Concepts are chosen as illuminating in part because they make possible the formulation and acceptance of principles that delimit mechanical, optical, energetic, biological and other possibilities in various domains. These principles are called 'nomological' generalizations. They include relativistic principles delimiting possible mechanical effects and experiments, optical principles such as the limiting speed of light, which similarly delimit possible optical and electromagnetic effects. Thermodynamic principles impose restrictions on possible heat engines and genetic principles restrict possible genetic processes, such as the heritability of acquired characteristics like mutilations.

Mechanical concepts generally play a role in a network of laws that link them with other concepts, which themselves similarly take part in a network of usually other laws. Mass is linked to Newton's first and second laws as a constant mediating force and acceleration and also as proportional to Newton's 'innate force of matter' (*vis insita* or *vis inertia*), which is manifest as inertial resistance to applied forces. Forces cause proportional accelerations by Newton's second law, can be decomposed or added as vectors and are generated by mass-pairs, charge-pairs, pairs of magnetic poles, etc. according to laws of gravitational, electrostatic, and magnetostatic force. 'Force' is not 'defined' by the second law in any usual sense of 'definition' (Mach, 1919) because not all forces actually cause accelerations and because the laws of vector composition and their application in statics are just as essential to the classical conception of mechanical force as are laws of motion. Thus, as Thagard and others suggest (Thagard, 1988, Holland, *et. al.* 1986), concepts like force should be understood by knowing their role in a default network of laws, that is, laws which hold presumptively but not always.

Change in law can result in changes in concept: The concept 'quantity of matter' preceded the Newtonian mass concept in Buridan's impetus theory as a measure of weight and resistance to movement. With the law of inertia mass becomes a measure of resistance to acceleration, and for the atomist Newton a measure of bulk in the sense of the number of fundamental atoms (*Principia*, Def. 1).

A second law-concept network can be found in species concepts. The essentialist species concept fits the Aristotelian eternalist theory of incorruptible natural kinds. According to this theory the regeneration of species is an image of eternal celestial cycles and represents the end or the perfection of terrestrial organisms, their final cause. This law-concept complex is also transferred to the creationist taxonomies of Ray and Linnaeus in the 17th and 18th centuries, who add just a Biblical beginning to the reproductive cycles of living species. An ecological component of species concepts dates from Linnaeus' 'economy of nature': Species are identified in part by their place in a nutritional ecological hierarchy organized to serve humans at the top. The geographical species concept appeared in the works of Buffon in the 18th century, where an attempt is made to explain why there are unique biogeographical regions such as South America, Australia, sharing African and Eurasian climatic zones but radically different in flora and fauna. For Buffon species are breeding populations confined to specifiable geographical regions. Lyell in the 19th century combines ecological with geographical attributes in his adaptation theory of species form and extinction. He also de-anthropomorphises ecology in his beliefs that biological and geological history is indefinite in extent and man is a recent occurrence. The contemporary biospecies concept adds genetic descent to the ecological and geographical characteristics of species. Distinct species occupy distinct historical lines of reproduction that are genetically isolated, and because of their genetic isolation they evolve independently of other genetic lines at least insofar as their genes determine their characteristics. This does not exclude ecological interaction of distinct species, where the presence of one species can influence another by influencing the selection pressures on that species. (Browne, 1983)

5. Theories or Theoretical Models T. Theories or theoretical models are products of attempts to generate explanations of particular entities, events processes and laws by applying concepts and laws to particular circumstances, that is as set of particular entities, their initial conditions and conditions at boundaries of the systems they compose.

A Newtonian model of the Sun and Mars includes a description of initial positions and momenta of these bodies, boundary assumptions about other bodies that might influence their motions, e.g. other planets might be excluded in a first approximation, and an application of relevant laws governing subsequent evolution of this two body system. What makes this a Newtonian model is that the bodies in the system are attributed Newtonian properties, location and momentum, and are taken to be subject to Newtonian laws; the momenta remain constant in time and the bodies interact gravitationally. In the model the combined effect of the laws will be a double orbiting motion of Mars and the Sun around a common center of gravity, and without external perturbation these bodies will trace two Keplerian ellipses. A second Newtonian planetary model might include Jupiter as a third body. In this case the boundary conditions are extended to include Jupiter's orbit in the system and Jupiter's influence on the system must be taken into account in calculating Mars' and the Suns' orbits. In this case there will be notable purturbations in the Keplerian characteristics of the planets' orbits, some of which will not occur in predictable cycles.

These models are means to testing Newton's theory of gravitation as well as practical devices for its application. It is the models that we expect to 'save the phenomena', that is to generate orbits that contain heliocentric longitudes, heliocentric distances and latitudes that are observed at various times for Mars and Jupiter. Since in theory, that is in all but the simplest model of the planets, the Sun itself orbits about a common center of mass shared with the planets, center of mass coordinates would be the most appropriate descriptions of the planets in the theoretical models. However, these coordinates differ in each model and the Sun's orbit is within the mass of the Sun and quite small relative to the dimensions of the planets' orbits. Hence heliocentric locations and motions for planets serve as the best available 'phenomena'.

In addition to 'saving the phenomena' we also expect the best newtonain models to cohere with background beliefs about the number and size of other planets. Thus a model containing Jupiter, Mars and the Sun can permit calculations of observed heliocentric coordinates that are of practical value for some purposes, but because background cosmology includes other massive planets in the Solar System such computations are descriptively incomplete and may not represent accurately all of the purturbations of Mars' and Jupiter's keplerian orbits. A realistic model will accommodate all of the planets within its boundary conditions but such a model is not necessarily a practical necessity. Because the several purturbations in realistic models may not occur in predictable cycles and their computation may be needlessly complex given the purposes for which the tables are constructed, less realistic one or two body models may be of more practical value in many instances.

We may also consider descriptions of species as models. In this case the entities under consideration are populations of interbreeding organisms. These organisms will be expected to have morphological characteristics that fall within a range of variation taken as normal for the species, including juvenile and sexual dimorphism, as well as behaviors and an ecological habitat of determinable geographical extent. A description of a species includes attribution to the population in question of values for these several morphological, behavioral and ecological characteristics. As do planets, species also contain dynamical characteristics, particularly those behaviors that give the organisms the power of occupying a geographical area within ecological boundaries, their powers of migration and dispersion, and morphological and behavioral characteristics which influence the organisms' reproductive abilities, their various adaptations and their survival or extinction. A species model, then, will consist of distinctive morphological and behavioral attributes of individuals in a population, their adaptive behaviors, organs and capabilities, and a description of their niche or ecological position, which would include a geographical region with ecological boundaries as well as parts played within the food chain of the region and associated behaviors and other characters. These are the items Darwin sought in his descriptions of Galapagos birds and plants as a means to answering the question about transmutation (Q1). It is in these terms that evolutionary events can be predicted, e.g. ecological or geographical isolation and expected changes in habit or form, genetic or population collapse and extinction.

It should be observed that formulating theoretical models can be as much a matter of conceptual inquiry as it is empirical inquiry into the applicability of received concepts and laws. In addition to identifying particulars in a domain to be explained, the formation of a suitable model can include the invention of new concepts that specify as well as instantiate Newtonian forces or that specify Darwinian adaptations and selection pressures. Such conceptual inquiries include efforts to obtain satisfactory quantitative measures of fitness and selection pressures. The line between 'normal' inquiry aimed at applying received principles and concepts and 'extraordinary' inquiry aimed at conceptual innovation is not as sharp as is suggested by some writers such as Kuhn, Lakatos and Laudan. Specific applications of a conceptual structure can require novel concepts, though generally these concepts are not incompatible with but compliant to the conceptual structures and beliefs in the paradigmatic background C, L and B.

6. Evidential relations R. Later defenders of the logical empiricist program (e.g. Hempel, 1954, Schaffner, 1969) acknowledged that 'correspondence rules' linking certain theoretical to observational terms are actually substantive in content or non-analytic. Defenders of the more recent 'semantic view' of scientific theories stress the role of auxiliary theories in establishing the required isomorphism between empirical and theoretical models (Thompson, 1989). However, it is also clear to most scientists that observation procedures and the functioning of instruments depend upon laws that link relatively reliably properties accessible to the senses with those not thus accessible, as the law of thermal expansion links temperature with the length of a column of mercury in the mercury thermometer.

It is because of the substantive character of evidential relations that Popper's 'logic of scientific discovery', viz. the elaboration of testable consequences by 'logical' deduction from hypotheses, is not at all strictly 'logical'. The testability of a scientific hypothesis is more likely to be a discovery based upon inquiries into the physics or chemistry of certain instruments than it is a product of logical analysis of theoretical terms. Also, as Feyerabend (1974) argues at length, one can discover by questioning received interpretations of scientific experiments, e.g. the Aristotelian interpretation of the 'tower experiment' in which the vertical path of a body in free fall is taken to falsify the hypothesis of the Earth's rotation.

7. The actualization of observable attributes E. In some cases causal relations between quantities reliably accessible to the senses and those not so accessible may be known without it being known how to implement the relations, say by manipulating initial or boundary conditions, so that the observable quantities are actually observed. Thus we might know that electrons give sufficiently short wavelengths to give clear images of objects the size of viruses without knowing how to get the electrons to focus on or reflect from the objects in question. In another case we might know that the wavelengths of certain X-rays are sufficiently short to diffract in crystals of a certain kind without being able to produce X-rays sufficiently close in frequency or sufficiently coherent to give a diffraction pattern from which the crystal structure could be computed. Thus the production of observables E known to be relevant through evidential relations R to some unobservable occurrence of interest is a problem generally independent of the discovery of R.

These several components of scientific cognition, the worldview B, metascientific beliefs M, available systems of concepts C,... and laws L,... with domains of application specified in B, and the evidential relations R and observables E are generally logically independent of one another, can be varied independently, and can become independently problematic. This latter point Feyerabend noted in his remark about the 'uneven development' of scientific belief systems (1974). However, generally when one element becomes problematic the others are not. Accordingly, those that remain unproblematic remain as the basis for formulating problems and selecting heuristics for investigations into the problematic components (Cf. Laudan, 1984). The following exemplify some general types of scientific problem, as determined by this epistemic taxonomy:

(i) *Historicist metascientific inquiry:* Given background B, concepts and laws C,L, a series of models T,T′,..., evidential relations R,R′,..., with actual and possible empirical evidence E,E′,..., determine methodological and heuristic principles M, viz. epistemic and other objectives and means for their achievement, whereby choices in the progressions T,T′,…, R,R′,…, E,E′,…. can be considered as rational, that is as a process of reasonably chosen ends and means for achieving them.

This is the fundamental problem in Lakatos' methodology (1970, 1971), which aims at drawing methodological principles from a 'reconstructed' history of science, that is, a history abstracted from irrational and 'external', e.g. social, influences. Lakatos has been criticized for placing too many Popperian constraints on efforts to solve this problem, particularly constraints demanding that 'rational' choices be progressive in that the succession of theories T,T′,… have associated ever richer evidential relations R,R′,… with ever more novel E,E′,…. Lakatos can be rightly accused of prejudging history and distorting it in reconstruction with this constraint. But these constraints need not be imposed in the pursuit of this problem, and non-Popperian values and methods may be sought in the history of science without the kind of reconstruction that Lakatos and his followers have in mind (Cf. Laudan, 1984).

(ii) *Logical positivist and logical empiricist inquiry:* Given a metascience M consisting of procedures derived from deductive logic and probability theory and of empiricist epistemic values, including deductions of evidential relations R from theoretical concepts C and a fixed empirical language E, elaborate further metascientific ideals for M compatible with logic and empiricist principles, such as metascientific conceptions of explanation, reduction, confirmation and theoretical plausibility (meaningfulness).

The procedure for addressing this problem in positivist programs is explicitly deductive and abstract from substantive theoretical or ontological commitment. Post positivist philosophers have gone to great lengths to show quite correctly that the presuppositions of this problem cannot be taken for granted, e.g. that E is not constant or universal, evidential relations cannot always be deduced from acceptable hypotheses, and first or second order predicate logic and probability theory are insufficient to account for all aspects of scientific reasoning. Accordingly, we can reject positivistic metascience as founded on false presuppositions but embrace historicist metascience without the Popperian constraints, and this is the program of this book.

(iii) *Scientific metaphysics:* Given some novel complex M,L,C,T,R, E (where T,R,E may include several models, evidential relations and empirical possibilities or actualities, a notational convention that will be observed below), find a system of background beliefs B or a revision B′ of existing B such that the complex B′,M,L,C,T,R,E forms a coherent whole. That is, find revisions in current world views that will incorporate coherent generalizations of this theoretical and conceptual novelty.

Such investigations have been carried out into the 'philosophical significance' of evolution by natural selection, quantum mechanics, special and general relativity, molecular and developmental genetics, cognitive ethology and machine intelligence (Capek, 1961). Plato, Aristotle, Kant, Reichenbach, Einstein, Heisenberg, Bohr, Capek, Bunge, Popper, Monod and Crick have all engaged in this kind of inquiry regarding the implications of geometry, biology, Newtonian physics, relativity, quantum mechanics, stochastic theories and molecular biology. A good deal of 'scientific metaphysics' about ontological hierarchies of composition and organization is taken for granted by writers in philosophy of biology and philosophy of mind (e.g. Hull, Maull, Wimsatt, Bechtel, McCauley and Churchland).

(iv) *Conceptual inquiry.* Given worldview B, or some prior adjustment B′ in received B, metascientific objectives M or some prior adjustment M′, find concepts C such that laws L can be formulated and a series of testable models T,R,E can be generated, or the generation of such series can be shown to be promising, where the series meet epistemic objectives of specificity, coherence with B and empirical adequacy prescribed in M or M′.

This kind of inquiry is exemplified in Darwin's research program in natural history and biography, where Darwin extends epistemic criteria from Newtonian methodology to these disciplines, particularly Newton's *vera causa* principle and the implicit simplicity of his planetary theory. Darwin then seeks concepts of variation, inheritance, genetic isolation and adaptation so that the occurrence of closely related but distinct species, both extant and fossil, in various regions of the world can be shown to be the outcome of a natural genetic or reproductive process in the circumstances prevailing.

In borrowing metascientific principles from physics and applying them in biogeography, the effort to explain the distribution of living and extinct species throughout the world, Darwin revised the then received metascientific objectives in natural history. These objectives include parts of and developments upon the creationist metascience of John Ray in the seventeenth and Carolus Linnaeus in the eighteenth, and Charles Lyell and William Whewell in the nineteenth centuries. Newton also influenced modern creationism in holding that the initial conditions under which the planets' motions are determined by gravity and inertia, viz. their existence, relative locations, masses, and illumination, are products of divine creation. Kant's and Laplace's hypotheses regarding the mechanical evolution of the solar system from a rotating gas cloud provided a theory of the lawlike evolution of mechanical conditions in the solar system, and in the eighteenth century this evolution was the basis for speculations by the French naturalist Buffon about the age of the Earth that extended geological time by a factor of 10^4. However, the apparent design of organic beings motivated natural historians to hold the thesis that each species was specially created for its ecological circumstances, though nineteenth century naturalists had acknowledged occasional extinction and had given up the view that all organic beings were created at once at the beginning of the Biblical time scale. Thus, though there are differences among creationist naturalists, one common element is the permission to use and positive valuation of the explanatory power of divine purpose, benevolence and creative action for physiological and ecological order. Accordingly, Darwin's borrowing of Newton's metascience is partial and selective. This point raises questions about how his partial borrowings can be rationally justified. In identifying humans as just another species of animal, thus rejecting the traditional exceptional biological status of humans, Darwin also modifies traditional background beliefs in a manner that significantly influences the direction of his inquiry.

Darwin exploits this identity in his willingness to infer inductively from characteristics of human reproduction, particularly human demography, to that of animals and plants, as well as in the converse direction. The identity provides the similarity requisite for him to consider the survival and extinction of human families to be an effect like the survival and extinction of animal and plant populations and to search for like causes among lack of fecundity, disease, starvation, and so on.

The same kind of problem is exemplified in the revolutionary research programs of J.B. Watson and Francis Crick. In this case B is given in the form of physicalistic, as opposed to vitalistic beliefs about reproductive, nutritive, behavioral and cognitive life processes. One relevant modification of B was made early in this century, the admission in chemistry of the possibility of very large molecules, of atomic weight 10^5 or more, among which were found organic macromolecules such as proteins and nucleic acids. Also several laws and conceptual frameworks, L,L′,..., C,C′,..., were given for the domains of quantum stereo-chemistry, which places constraints on the spatial configuration of various chemical bonds, the biochemistry of macromolecules, which permits large molecular strings and spirals and distinguishes various substances such as proteins and nucleic acids, and the crystallography of organic compounds in which methods were developed for constraining crystal structures by X-ray diffraction experiments. Further constraints come from the genetics and parasitology and pathology of viruses, bacteria and micro-fungi, including the thesis that DNA is the genetic substance, the hypothesis that genetic differences and genetic specificity can be accounted for by different distributions of four nucleotides on DNA, and associations of genetic differences with the production of specific enzymes. Other constraints come from genetics itself: genetic material must be capable of self replication, dissociation and conjugation in order to be heritable according to Mendelian principles. Watson's and Crick's aim was to take these conceptual and experimental constraints and obtain a conception of the structure of DNA that would also contribute to explaining its genetic characteristics, e.g. its genetic specificity and self replication.

(v) *'Normal' scientific inquiry.* Kuhn's normal science takes as given a worldview B, concepts and laws C and L, and certain largely implicit metascientific values M, experimental procedures involving presumed evidential relations R and an empirical language E using concepts from C. What is sought are theories or models T, which are specifications of C for further components of the subject matter defined in B, and the production of actual instances of E such that the propositions of T match them.

Lakatos' variant of Kuhn's normal science acknowledges that evidential relations R number among the auxiliary assumptions needed in the development of a model T, and the adds certain explicit 'negative' and 'positive heuristics' to M as well as Popperian demands for 'empirical progress'.

Kuhn has been criticized for holding that the paradigm, a complex of ontology B, concepts C and Land E, metascience M, must be viewed holistically (Laudan, 1984). B,C,L and E are allegedly semantically holistic in the sense that meanings, the concepts C and E, are determined by ontological and nomological beliefs in B and L, and any revisions B′ or L′ produces complexes B′,L′,C′ and E′ that are wholly disjoint from the original components. For empiricists this renders paradigms methodologically as well as semantically incommensurable. No common M can be appealed to in appraising two or more paradigms, and differences in C infect both B and E so that no two distinct paradigms have a common subject matter.

Now if methodological ideals cannot be separated from paradigms then revolutionary or 'extraordinary' science is inevitably a-heuristic and non-rational. It can contain no epistemic objectives and can entertain no means for achieving them. Thus, if extraordinary science entails that all previous scientific commitments be abandoned, as implicit in the radically holistic interpretation of Kuhn's position, then nothing recognizable as scientific inquiry would be possible. Metascientific inquiry does not necessarily presuppose metascientific objectives, but it does presuppose consensus on scientific achievement upon which metascientific proposals can be modeled and authenticated. Ontological inquiries similarly presuppose exemplary scientific achievements.

Conceptual inquiry, as described above, presupposes metascientific objectives and ontological and empirical commitments to events and processes whose explanation is sought. Thus we cannot construe ontological, metascientific, conceptual and empirical commitments to be so organically related that they cannot be separated to ground revolutionary inquiries, particularly what have been described above as metascientific, ontological and conceptual inquiries.

The continued possibility of scientific inquiry can be compared to the continued possibility of scientific progress and scientific discourse as regulative conditions placed upon philosophical analyses of science. Philosophical epistemologies must give some historically and intuitively satisfactory account of the progress of science, philosophical methodologies must give similarly satisfactory accounts of the possibility of scientific inquiry at all stages of its history, and theories of scientific language must maintain the possibility of understanding views commonly discussed in the history of science and philosophy. Otherwise such views cannot cite history for evidence.

2.5. SUMMARY

Scientific discoveries combine the invention of concepts with novelties in scientific encounters with nature. In turn these encounters can be produced by novel developments in auxiliary theories which provide relations of empirical evidence or in novel technologies for manipulating, observing and measuring components of the world. Ontological, conceptual, metascientific and empirical components of scientific belief systems are distinguished as independent variables in scientific development and inquiry. These components can be compared with Kitcher's [1985, 1990] concept of a *practice*, which consists (i) of a language (our law-concept complex, C,L) (ii) a set of statements (theories and theoretical models T, laws L and observational statements E) (iii) a set of questions (iv) a set of patterns of reasoning (heuristic principles in M) (iv) a set of experimental techniques (evidential relations R and heuristic principles in M) (v) a set of methodological directives (other metascientific prescriptions in M). As to Kitcher's (iii), various kinds of problem or question in science, conceptual, metascientific, ontological and empirical, are defined as perceived deficiencies in these various components of a practice.

CHAPTER THREE

EROTETIC LOGIC AND SCIENTIFIC QUESTIONS

3.1. INTRODUCTION

Science is commonly called an institution and an instrument for solving problems, and sometimes it is said that some or all of scientific achievement can be reduced to finding the right problems or formulating scientific questions correctly or perspicaciously. We may acknowledge the importance of proper selection and formulation of problems without accepting all of this reductionistic thesis because the pursuit of problems, that is the search for their solutions, is also a large part of science. Methods and guidelines for the pursuit of science as well as scientific problems should count as much as a part of scientific knowledge as do scientifically vindicated beliefs about various parts and aspects of the Universe. This point does not exclude the role that scientific knowledge has in supporting scientific methods.

Until recently few philosophers have paid much attention to scientific problems, the knowledge of scientific problems, their choice and their pursuit, or knowledge of methods for their pursuit. This chapter and the next is an attempt to partially fill this gap in our philosophic knowledge of science. This chapter will focus on what are scientific questions, the next will focus on the methodology of choosing or asking scientific questions, and the chapter to follow will focus on answering them, that is the search for answers to scientific questions.

Specifically this chapter will include a summary examination of *erotetic logic*, or the logic of questions. Here we shall consider what illumination erotetic logic offers for a satisfactory conception of scientific questions and problemshifts that take place in the history of various scientific fields. We shall seek clarification of what questions are, and whether there is anything distinctive about scientific as opposed to other kinds of question. In chapters to follow we shall consider what constitutes knowledge of questions, the assessment of questions for their heuristic merit, and what considerations underlie the choice or the preference of questions in designing programs of scientific research.

An exhaustive survey of all that is written on erotetic logic or the various ways in which erotetic logicians construe questions will not be attempted here. Rather one such construal which seems particularly promising as a logic of scientific questions will be considered. A suggestion as to what this promise is will be offered and appraised as a source of illumination for the methodology of scientific inquiry. Our conclusions regarding this system of erotetic logic will, in turn, apply *a fortiori* to most other formulations as well.

3.2. QUESTIONS IN AQVIST'S EROTETIC LOGIC

Lenert Aqvist's erotetic logic (1975) combines *epistemic* logic (a logic of belief and knowledge) and *deontic* logic (a logic of obligation and permission). Hintikka takes a similar approach in his writing on questions (1976, 1981). Aqvist's and Hintikka's logic of questions is based upon the assumption that questions are best paraphrased as requests or demands for knowledge.

It might be thought that our aspirations to knowledge have varying degrees of importance, and thus our obligation to seek a particular item has varying degrees that cannot be captured in ordinary deontic logic. However, a distinction must be drawn between the content of a question, which includes a request or a demand, and the appraisal of the question which can be made in the light of various values attached to the item of knowledge. According to its appraisal, a question may provide varying degrees of obligation to provide or search for the knowledge that the question demands. More will be said about the appraisal of questions in the next chapter.

This construal of questions as expressing aspirations to *knowledge* is *prima facie* plausible because scientific objectives include states of belief and knowledge and whatever is presupposed by such states in the way of conceptualization and certifiable evidential relations. Also, intellectual history in general and the history of science in particular seek ascriptions of beliefs and knowledge to individuals and to communities, particularly members of various scientific disciplines at various times. Proper historiography in these disciplines requires that anachronisms be avoided, including anachronistic attributions of modes of conception as well as objectives and capabilities of inquiry. The intensional apparatus of epistemic logic distinguishes between, e.g., ascribing to Aristotle the belief that the Earth is spherical, which is appropriate to Aristotle's cosmology, and ascribing to Aristotle the extensionally equivalent belief that the third planet is spherical, which is conceptually anachronistic. Hence an adequate logic for scientific questions should have a component that will capture epistemic components of scientific inquiry, particularly the intensionality of the questions asked and statements made by scientists of a given conceptual community. Also one might expect logical differences between the intensional modalities, knowledge and belief, that will be useful in illuminating general features of epistemic inquiry.

Furthermore, the epistemic component of erotetic logic should aid in accounting for how scientists are capable of believing or knowing unfulfilled scientific objectives, particularly epistemic objectives. If the goals of human activity can include epistemic goals, that is, goals whose fulfillment is or can be counted as knowledge, if these goals can be known independently of their fulfillment, and if there are reasoned valuations that bear upon the preferential ordering or choice of these goals, then rational inquiry, at least in the sense of rationally posing questions, is possible. Accordingly, a logic appropriate to scientific inquiry must have some way of expressing epistemic goals or ideals that are not, but can or should be actualized. It should also have the power of expressing and perhaps even certifying some, if not all the conditions under which these goals are or can be valued, chosen or preferred.

Aqvist's logic is capable of expressing quantified as well as unquantified questions. Here we shall focus on the quantified part (quantified imperative epistemic logic, or QIE), and we shall discuss the propositional part (propositional imperative epistemic logic, PIE) only insofar as it is conducive to simple and illuminating formulation of important general principles applying to both parts.

3.3. QUANTIFIED IMPERATIVE EPISTEMIC LOGIC (QIE)

A simple and clear account of partial and cumulative information about any subject can be given in terms of standard predicate logic, whether that subject be an individual person or object, an event or a process. This information is provided by ascribing properties and relations to these individuals, and is readily extended, by the inclusion of number theory in a conceptual background, to ascribing quantities of various kinds, including scalars and vectors of any number of dimensions. Greater descriptive power is obtained by adding to the language second order properties and relations that might be borne by the properties, relations and quantities describing individuals. For example, one might describe a property as extensive or intensive, or presently observable. A quantity may be described as observable within a range of its values, or as a vector with specified dimensionality, and one might describe a relation as isomorphic, homeomorphic, reflexive, symmetric or transitive.

Another means of extending the descriptive power of predicate logic requires some reflection upon names and other singular terms and the ontological status of various items that can be designated by these terms. These considerations are also necessary for choosing appropriate interpretations of the expressions appearing in QIE. With explicit and appropriate background beliefs regarding identity conditions, particularly regarding spatiotemporal relations and the tokening of concept types, but also concerning historicity of concepts, one may construe concepts as namable abstract individuals with spatiotemporally localizable tokens. Properties and relations, such as observability, natural, qualitative, quantitative, relational, scalar, vector, real-valued can be ascribed to concepts. Individuals may be broadly defined as whatever can be individuated, that is whatever can be described uniquely or distinguished by other means such as ostention from other individuals of greater or lesser similarity. Concepts can count as individuals under this definition.

Numbers may be considered paradigmatic concepts in the effort to determine how concepts can be identified and named. Numbers are, in contemporary epistemic contexts, construed as sets of equinumerous sets and can be identified by beginning with the set of empty sets (zero) and successively applying the successor relation for identifying the set of uniquely occupied sets, the set of binarially occupied sets, etc. Among the properties and relations of numbers are 'even', 'odd', 'prime', 'equal to', 'greater than', 'rational', 'real' and 'imaginary'.

Turning to scientific concepts, species, such as Darwin's Charles Island mockingbird, *Orpheus Trifasciatus*, are in a sense concrete individuals. They are synchronous populations of relatively recently genetically related sexual and asexual organisms that, in the case of sexual organisms, are potentially interbreeding (Cf. Hull, 1985, esp. Chap. 5). They are also in this sense components of a phylogeny, a temporally extended population of organisms among which mutually genetically isolated subpopulations of recently genetically related organisms generally increase in number with the passage of time. In another sense 'species' is a concept, a way of dividing and describing by classifying individual organisms according to their anatomical structures, behaviors, geography and genetic relations.

By comparison, number concepts presuppose a division of the world into individuals and aggregates and classifying the aggregates according to their equinumerality. Unlike numbers, species have a history that is more than just a bounded temporal extension, for the characteristics that are distinctive of a species are always subject to change, though individual species are identified by families of attributes that are constant for small spatiotemporal intervals. Whereas numbers are generally considered timeless in at least the sense that their properties do not evolve, still number concepts had a beginning when humans began first to count. Thus numbers cannot reasonably be considered timeless Platonic entities that precede all intelligence.

Still both numbers and species can be named. The names function in producing identifying references in communication when the communicants, e.g. the questioner and the provider of an answer, share background beliefs, such as those just described for numbers and species, pertaining to the individuation of the named entities. These background beliefs include conceptions or beliefs about what it is to know, respectively, numbers and species, at least within a given epistemic context.

As in taxonomy, with appropriate care regarding what is presupposed in naming, particularly sufficient descriptive knowledge for identification, concepts, like individual persons Richard Nixon or species *Nesomimus Trifasciatus* (Charles Island mockingbirds), can be designated and referred to by indices, that is, labels that are non-descriptive for a given language community. Thus the English word "Force" or the symbol 'F' can designate without in any way describing concepts that can be distinguished and identified in the history of science. Aristotle's conception of efficient cause, or (exclusively) one of the various of Newton's conceptions of intrinsic and extrinsic activities and powers, can, in the appropriate contexts, be designated by "force". "Force" is not representative or descriptive of the structure or content of these concepts, and thus Feyerabend's and Kuhn's meaning variance thesis, viz. that "force" can have various meanings in various historical contexts, is a possibility. The name "Richard Nixon" represents identifying characteristics of an individual person only by having associated with it in the minds of members of a language community membership in the Republican Party and the condition of being the first president to resign in office, possibly among other attributes. However, nothing in the token here or the name type that, apart from associations in many politically knowledgeable individuals' minds, represents these identifying characteristics. "Charles Island Mockingbird" indicates that designating names or nouns, in this case a noun phrase, can be partially descriptive for a given language community. In this case it is descriptive of the habitat of the mockingbird species designated by the phrase for geographically knowledgeable English speakers. However, species designators can also be indexical, as the Latin name *Nesomimus Trifasciatus* demonstrates. Individuating and distinguishing properties, properties that collectively or individually uniquely describe or distinguish an individual from other like individuals, are equated with or attributed to whatever individual the name indexes in an effort to conceive this individual human, species, or whatever.

Identifying an individual, or the species to which an individual belongs, is a matter of finding a sufficient number of properties that describe the individual or species essentially (without which that individual would not be what it is) and uniquely (properties that only the individual in question has). This point does not entail traditional essentialism when it is understood that the cluster of 'essential' characteristics is relative to a conceptualization, and conceptualizations can differ among communities or change in time. The conditions for sufficiency of such identifying descriptions can be various for various purposes, disciplines or historical stages in those disciplines. One set might be sufficient for a freshman philosophy student's identification of Aristotle, but hardly so for the identifying description that an Aristotelian scholar might demand or offer. Such conditions can be stated in the form of metascientific desiderata M, for, e.g., professional historical scholarship in philosophy.

Causal theories of reference confuse the act of establishing a nominal convention, on the one hand, e.g. that "Force" in some context of discourse will designate only newtonian applied accelerative force, or that "Richard Nixon" will designate only a person (an infant) meeting a certain cluster of individuating conditions, with various acts of reference on the other. Such acts include, e.g., referring in fictional discourse to Santa Claus or in intellectual history to Aristotle's efficient causes. Neither act is an intensional fiction. The acts of reference are actual once performed and may be considered successful if they elicit in others attention to the same individual, whether or not that individual referent is fictitious. The attempt to establish the existence of gravitons is an effort to show that an intensional actuality, the reference to gravitons, corresponds to a physical actuality, the gravitons themselves. In the last sentence I presume I have just succeeded in referring to gravitons without establishing that they exist. The former task is in the competence of reasonably well-informed readers of popular science when they address equally well-informed audiences, but the latter task is in the competence only of certain experimental physicists, if even that at present. This last effort, the attempt to show the actuality of a referent in fundamental physics, involves various acts of conceptualization, and the construction and preparation of instruments and careful measurement, and is directed to showing not that statements in which "graviton" occurs transparently actually refer, but to showing that they refer to something actual.

With these preliminaries, let us now consider the sentences:

> Someone in this room was born in Alaska.
>
> Something is a native of Charles Island and is a species of the genus *Nesomimus*.

As commonly understood, these sentences describe a person born in Alaska and a species of bird native to Charles Island without disclosing the identity of either one. That is, they provide partial information regarding two individuals that are not identified in the descriptions. It is a simple matter of conjunction to enhance this information:

> Someone in this room was born in Alaska and now lives in California.
>
> Something is a native only of Charles Island and is a species of the genus *Nesomimus*.

Also these sentences can readily be made less informative:

> Someone in this room was born in either Nevada or Alaska.
>
> Something is a native of either Charles Island or Albemarle Island and is a species of the genus *Nesomimus*.

Thus we can enhance or weaken information about an undisclosed individual without revealing that individual's identity. Enhancing such information is one very effective method for discovering the individual's identity, and this method meets all of the requirements for a heuristic described above.

The next step in developing a formulation for a question is to express the goal that the objects described be identified, that is that they be known to have sufficient identifying descriptions. An individual that is credibly (N.B. 'credible' means 'believable' or 'acceptable as true') described by sufficient identifying description counts as an individual that is known. 'Sufficient identifying descriptions' are not here intended to exclude ostentions, which are, in some epistemic contexts, fully sufficient for identification. Ostentions can be, though they need not be, expressed with a kind of description, 'This person', 'This bird', and even 'This kind of bird'. Pointing and grunting can in some contexts succeed as an ostention, and it is stretching a point to call this act a description. Still in certain contexts of shared background beliefs such actions can answer questions and constitute sufficient identifications of individuals in question, though the actions are only a small portion of human powers of identification.

There are two components in expressing epistemic goals of identification: (i) that some objective be actualized and (ii) that this objective be an identification, in the above cases either of a person or of a species of mockingbird. The identification sought is also to be an item of knowledge, a sufficient number of individuating propositions that can be accepted as true with sufficient certified evidence. At this point we need some logically suitable expression for what is sought, the desideratum of the question. In both examples *nominal* and *informative* answers need to be distinguished (Belknap and Steel, 1976). "Tom Smith" and "*Nesomimus Trifasciatus*" are terms in common English and common taxonomese of the 20th century for persons and genus-species complexes respectively, but this information is insufficient for providing desired answers in any typical usage of the above questions. To answer these questions one must also provide some knowledge of who Tom Smith is and what is the species *N. Trifasciatus*, that is, the person Tom Smith and the species *N. Trifasciatus* must be identified in answering this question. To identify a person or a species is to know that a sufficient individuating description applies to them, that is to believe with sufficient justification such descriptions or to know them by such descriptions.

In the case of persons, ostentions and individuating biographical descriptions of varying degrees of detail, depending upon the kind of inquiry undertaken, will count as sufficient in answering questions like "Who in this room was born in Alaska?" or "Who was Aristotle?" One such inquiry might be the introduction of students to one another in a classroom with the objective of producing a cooperative group for pedagogical purposes, where it is supposed that ıshared information about geographical distribution is conducive to this end. In this case ostensive answers may be sufficient. In another kind of inquiry, viz. the teaching of the history of philosophy, geographical and historical identifying descriptions are insufficient and ostensions are impossible in the effort to identify Aristotle. Instead one would seek identifications by attributing works and philosophical views some of which, or some combination of which, would be idiosyncratic.

In the case of identifying species such as mockingbirds, taxonomy has rather specific standards, though in some respects these are subject to controversy, and they have changed, as illustrated by the Darwin case study in the previous chapter. These standards are embodied in a current species concept, and as such they reflect received beliefs about how much variation in appearance, deep morphology, and behavior among juveniles, males and females can be expected within an interbreeding population or a biological species. The species concepts thus formulated may also be generalized and thus considered standards for appraising various species identifications made by various naturalists, in which case they fall also under metascientific prescriptions M for good species identifications. Questions of species identification may be schematized as follows: Given geographical and other background knowledge, taxonomical concepts C appropriate to a set of laws L (the various laws governing evolution, particularly those bearing on breeding within a population), and evidential relations R appropriate to circumstances prevailing in some geographical locality (relevance of morphology, behavior, habitat to genetic isolation), produce some observations E (appearance, morphology, habit, song) such that a specific description T instantiating C and satisfying metascientific desiderata M for specificity and empirical adequacy is obtained. It should be noted that in facing his problem of deciding the specificity of the Galapagos mockingbirds, Darwin called into question his species concept C as well as sought specimens describable by terms in E that would enable him to solve this problem.

Specific conceptual systems C,L, metascientific objectives M and evidential capabilities R,E constitute a *context of inquiry* in which theoretical and observational terms appearing in both questions and answers will have more or less specific meanings which can but need not be peculiar to that context. This point, contrary to the thesis of radical meaning variance, does not exclude the possibility that other terms may share the same meanings in several contexts. Also meanings of questions or other components of a context of inquiry can embody research objectives as well as beliefs about a subject that can differ from those in other contexts of inquiry. In the same context possible answers can meet or fall short of these objectives as seen from the viewpoints embodied in these meanings. The objectives that may be distinctive to a context of inquiry will include or presuppose metascientific standards that determine sufficiency in identification and knowledge of answers to the questions arising in that context, and we have already seen how these can differ from one context to another. Accordingly, the language utilized in a logic of questions that might be useful for the formulation of scientific questions must be sufficiently rich to allow the expression of differences and similarities among the various contexts of inquiry that occur in the history of science.

In epistemic logic the propositional operator "K" in the expression "Kp" is taken as an expression for the proposition that p is known. K can be indexed to individual knowers *i*, as in K<*i*>, so that "K<d>p" can be read "Darwin knows p". Various interpretations might be given to "K<*i*>p", but these should be constrained by the axiom in epistemic logic:

$$(\mathrm{K1}) \vdash \mathrm{K}\langle i\rangle \mathrm{p} \rightarrow \mathrm{p}$$

In plain English, if person *i* knows p, then p. K1 is a more formal expression of the commonsense maxim that only true propositions can be known. Hence we might be willing to ascribe to Darwin knowledge of the geographic range of *Nesomimus Trifasciatus* if he believes, in agreement with us and with evidence we are willing to certify, that the bird is confined to Charles Island. However, Darwin did not know this species by the name "*Nesomimus*", for he used the generic name "*Orpheus*" for mockingbirds. Because of the intensionality of "K", the first occurrence of "p" in K1 must be in Darwinian terms, but the second need not be. Rather, the consequent of K1 may be expressed in any terms that convey the meaning of "p" in the antecedent. That is, we can judge Darwin's claims to knowledge if we can paraphrase in our own terms and concepts what he sought in his own terms and concepts. We should thus deny him knowledge of the propositions of his pangenetic theory, according to which body parts transmit information to germ-cells by means of certain particles, what we understand as the sense of Darwin's "gemmules". There are no gemmules in the actual world, as we currently think we know it.

Under this interpretation of the operator "K" knowledge ascriptions to historical figures must be in the terminology and conceptualization of the historical person but their acceptability is not relative to the context of inquiry in which that figure operated. We thus reject the extreme form of the 'principle of charity' according to which any judgment of another's belief to be false is evidence that the other's terminology has not been properly understood.

The *epistemic setting*, that is, the ontology, metascience, conceptual systems and empirical capabilities, of present day science is implicit in the acceptance of p, or its contemporary paraphrase, as part of the actual world. However p is expressed, if it is acceptable it must cohere with the epistemic setting of present day biology. Though Darwin's belief in pangenesis may have been justified in the epistemic setting prevailing for naturalists and geneticists in the 1860's, we can only ascribe him belief, $\mathrm{B}\langle \mathrm{d}\rangle \mathrm{q}$, of the propositions of pangenesis q. There is no principle for $\mathrm{B}\langle i\rangle$ corresponding to K1 in epistemic logic. However, belief ascriptions, like knowledge ascriptions, are intensional. Accordingly it is correct to ascribe to Darwin the belief that *Orpheus Trifasciatus* uniquely inhabits Charles Island, but it is incorrect to ascribe him the belief that *Nesomimus Trifasciatus* inhabits that island, for Darwin used John Gould's ornithological terminology and not the more recent terminology that includes '*Nesomimus*'. Also it is correct to ascribe to Darwin the belief that certain characters dominate others in hybridization, but it is incorrect to ascribe him the belief that certain genes dominate others, not just because "gene" appears only in 20th century terminology, but also because Darwin never conceived of anything comparable to Mendel's heritable 'factors', which, because they can be recessive, must be distinct from the characters they determine. Intensionality thus requires historical specificity in both conception and in terminology in ascribing belief, but such specificity is not required in ascribing knowledge.

Questions in QIE can express *aspirations* to knowledge, which are distinct from actual knowledge. In QIE the posing of a question is the hypostatization of an ideal world, called an *ideal epistemic world*, not an actual world, in which a certain knowledge state exists. 'Worlds' are sets of *models*, i.e. sets of possible realizations of various sets of propositions. A Euclidean world is a set of possible realizations of formal euclidean axioms, which may include sets of real numbers but, importantly, also a physical space with zero intrinsic curvature. Aqvist's conception of semantics for questions is this: An epistemic alternative world is identified as a world distinct from the actual world. The actual world contains some knowledge KP and corresponding facts P, but the alternative world is distinguished only by its containing knowledge of some proposition Kq. Only Kq, and consequently also q in the alternative world is not also part of the actual world. The alternative world contains also KP and P found in the actual world. An ideal epistemic alternative world is an an epistemic alternative world whose actualization is demanded. A *question*, e.g. 'Is q true?', demands the actualization of an epistemic ideal world, viz. one distinguished by the knowledge of q or one distinguished by the knowledge of not-q. In the case of quantified questions such as

'What species of mockingbird is endemic to Charles Island?' there can be an indefinite or various infinite numbers of ideal epistemic alternative worlds, one for each possible answer to the question. All that is actually known about these worlds is that they contain actual knowledge and some as yet unidentified endemic mockingbirds on Charles Island.

Now it is possible to ascribe to Darwin aspiration to knowledge of gemmules or certain features thereof without ascribing him actual knowledge of such. The latter ascription commits us to the existence of gemmules or to an actual world with gemmules having various properties, relations, etc. Ascribing epistemic aspirations to Darwin can be done contextually, i.e. in view of the context of genetic inquiry prevailing and available to Darwin in the 1860's. All we need to say is that Darwin hypostatized an ideal world like the world he thought he lived in every respect except that in that ideal world he is in possession of, e.g., knowledge that gemmules determine gender. Actually in asking his question 'Do gemmules determine gender' he hypostatizes two ideal worlds, one in which he knows they do and one in which he knows they do not determine gender. The first of these worlds is distinguished from Darwin's background ontology not just by its containing the belief that gemmules determine sex, but also by that belief's satisfying then prevailing metascientific, methodological and empirical criteria for knowledge. Thus Darwin is aspiring to knowledge and not just cognition or conception of this feature of gemmules. There are gemmules in Darwin's hypostatized worlds, but there need be none such in the actual world. If the first of Darwin's aspirations is realized, his *weltanschauung* will be supplemented by the addition of sex-determining gemmules. Also, if we ascribe to him the realization of this aspirations, our *weltanschauung* will also contain sex-determining gemmules. Only then will we believe that gemmules occupy the actual world. However, such content would render our worldview incoherent because of the presence therein of the fact that inheritance is not influenced by acquired characters. The actual world as we take it cannot include gemmules. Thus Darwin's aspirations to knowledge of gemmules are unrealizable in the actual world in our present day epistemic setting. Darwin's question does not meet the requirement of compatibility with what we take to be fact, and hence, even though Darwin thought he was justified in asking this question, it is, in our view, a *false question*. Pre-analytically, the falsity of the question is apparent from our belief that there are no gemmules. Asking about the properties of gemmules presupposes that there are gemmules, but this is a false presupposition, so we cannot legitimately ask for properties of gemmules. We can still attribute to Darwin the aspiration to this knowledge, and since these attributions are intensional, they must contain the correct historical terminology and concepts. It is one thing to attribute aspiration to knowledge of a certain kind and quite another to appraise this aspiration as correct or justified in our own worldview, metascience and conceptual system.

Indexing the operator "K" to individuals is essential in attributing knowledge because the intensionality of different individuals differs, particularly when they participate in different contexts of inquiry. Thus we may stipulate that understanding the index, e.g. understanding the difference between "Darwin aspired to knowledge of inheritance" and "Mendel aspired to knowledge of inheritance", includes cognition of differences of conception and terminology in the thought and expression of Darwin and Mendel. Accordingly, the proposition "K<d>q was aspired to" should be read as ascribing an aspiration to knowledge of q to Darwin, where these aspirations include or imply the epistemic and terminological setting of inquiry in the 1860's as Darwin is cognizant of it. Thus the aspiration to K<d>q is aspiration to peculiarly Darwinian knowledge. Some generalization of the index, e.g. universally quantifying it over a community of human beings, indicates that all members of this community of inquirers shared this particular aspiration. Such general ascriptions of questions suggests, but does not imply, that this community shares a common epistemic setting at least insofar as it is relevant to the knowledge aspiration attributed to them. If the knowledge sought is knowledge of something that explains an item of curiosity, an *explanandum*, then ascription of aspiration to such knowledge to one individual can occur in pedagogical contexts in which one individual seeks explanatory knowledge shared by a community, say a scientific community that constitutes the membership of some discipline, such as population genetics. That is, 'Why-questions' ascribed just to individuals can express aspirations of individuals to enlightenment already available to a community. Generalizing the index in the same why-question to all members of an intellectual community is a way of expressing the aspiration to explanatory knowledge that the entire community lacks. Finally some who aspire to knowledge may do so incoherently with her own *weltanschauung*. We can attribute aspiration to knowledge to individuals or communities even when such knowledge cannot be justified within that community's epistemic context.

One may act directly or act to bring about actions on the part of others as a means to fulfilling epistemic as well as other aspirations. Normative guidelines for such actions constitute one part of 'logic' of discovery, the other part being the choice of the epistemic objectives themselves. Among such actions are the formulation and expression or communication of these objectives and the demand that actions be taken to their fulfillment. Questions convey both the objective and the demand for fulfillment in the form of a request with specific content. The request portion of the question is expressed by means of a deontic operator '!' which can be read in English as 'Let it be that...', 'Bring it about that...', or by other English expressions for imperatives. Thus

$$!(K<d>p \vee K<d>{\sim}p)$$

might express a demand that Darwin might make that either knowledge that gemmules determine sex or knowledge that they do not determine sex be actualized, in short 'Do gemmules determine sex?'

In the terminology just introduced the sample questions concerning Alaska natives and mockingbirds can be expressed thus:

$$!(Ex)K<i>(Px\&Ax)$$
$$!(Ex)K<d>(Sx\&Cx)$$

where "$K<i>$" and "$K<d>$" respectively stand for "person i knows..." and "Darwin knows...", "Px" and "Ax" for "x is a person in this room" and "x was born in Alaska", and "Sx" and "Cx" for "x is a species of the genus *Orpheus*" and "x is a native just of Charles Island". The expressions inside the operator '!' are what Hintikka (1981) calls *desiderata* of the questions, and they can be regarded as expressions that, when suitably interpreted, identify particular questions. The desideratum expresses the content of the request in the question. Different questions are thus capable of being expressed in terms of the different expressions composing the descriptive clauses following the epistemic operator. This is exactly the means for identifying propositions in standard predicate logic.

Thus, an important consequence of this formulation of erotetic logic is that conditions for identifying and differentiating questions are no different from conditions for identifying and differentiating propositions. Problem epistemologies or definitions of scientific fields in terms of sets of problems pose no special problems regarding semantic incommensurability. Problems of incommensurability include those in the interpretation of foreign conceptual systems and in their comparison in reference to a common set of metascientific beliefs or empirical beliefs for the purpose of epistemic appraisal. We are no worse off in the face of possible incommensurability problems in offering metascientific concepts defined in terms of questions than we are in defining them in terms of propositions.

3.4. ANSWERS AND PRESUPPOSITIONS TO QUESTIONS

Another concept preanalytically associated with questions is that of *answers* to questions. One problem we have set for ourselves is the clarification of what it is to know a question. Presumably there must be some distinction between knowledge of questions and knowledge of answers to questions, for otherwise we could not reasonably ask unanswered questions, and accordingly rational inquiry would not be possible. We also seek to clarify what, if any, heuristic value knowledge of a question has in the search for its answer(s).

It is often said that a question is not properly formulated or understood until it is clear what would constitute an answer to it. Other writers on questions have held that a question is equivalent to its possible answers, so knowing a question is knowing possible answers, but perhaps not knowing which possible answer is true. Thus 'whether-questions' or 'yes-no' questions are sometimes paraphrased as alternatives that constitute possible answers: "Is *O. Trifasciatus* a distinct species endemic to Charles Island?" might be paraphrased as "*O. Trifasciatus* is a distinct species endemic to Charles Island or *O. Trifasciatus* is not a distinct species endemic to Charles Island." However, the disjunction needs some expression indicating the desire or obligation to affirm one of the alternatives as opposed to just affirming the alternative. Perhaps something further might be added to indicate what is involved in an affirmation, e.g. that what is desired is that the affirmation be coherent with background beliefs, both observational and non-observational, that is, that it be an item of knowledge. Sayings about the equivalence of question and possible answer do not make it sufficiently clear how it is possible for one to know, or what is implied in knowing unanswered questions. They also do not indicate with any clarity whether *conceptual inquiry* is possible: In conceptual inquiry one should be able to know a question in the sense of having at least appropriate concepts for its formulation and for reasoned belief of its presupposition, with possibly further information bearing upon the question's scientific importance, and yet lack the concepts required for its answers. In such a case knowing questions cannot in any clear way reduce to knowing possible answers, though it might imply knowing something about possible answers. Conceptual inquiry may raise special problems because it would appear that the epistemic setting of the question differs from that of the answer. Since we have interpreted the index in the "K<*i*>" operators as referring to individuals or communities defined by specific epistemic settings, it may be inappropriate to ascribe the same indices as they appear in conceptual questions and answers thereto.

Logicians have distinguished a variety of different kinds of answers (Aqvist, 1975, Chapter VI), of which for our purposes an elementary formulation of *direct answers* will suffice. Intuitively, "Tom Smith", or "Tom Smith is in this room and was born in

Alaska" or "*O. Trifasciatus*", or "*O. Trifasciatus* is a species of mockingbird and lives just on Charles Island", respectively answer the questions discussed above.

In order that questions remain logically distinct from their answers, erotetic logicians generally stipulate that the direct answer A be *just* logically sufficient for inferring the question Q, i.e. for inferring Q's desideratum, Core(Q) (Aqvist, 1975, p. 141). If A were part of a larger propositional context that entails Core(Q) and A also entails Core(Q), then A would not be just sufficient, but would still be a *sufficient answer*, not a direct answer to Q. The converse relation, viz. that Core(Q) entails A, is excluded by the possibility of inquiry, viz. the possibility of constraining and rationally defending unanswered questions. If Core(Q) entailed A, no search for A beyond logical analysis of Q would be necessary.

A constraint in choosing expressions for formulating answers to questions expressed above is that this condition for 'direct answer' be satisfied. Supposing that "s" stands for "Tom Smith" and "ot" for "*O. Trifasciatus*", "Ps&As" and "Sot&Cot" might *prima facie* appear to be respectively answers to the above questions. However neither of these propositions entail the respective desiderata (Ex)K<*i*>(Px&Sx) and (Ex)K<d>(Sx&Cx) because the existential quantifier in these expressions occurs outside the epistemic operator "K". Were the existential quantifier inside the operator, the expressions would read, e.g., "Darwin knows that something is a mockingbird and native just to Charles Island". This is not an expression of Darwin's desire to know *what* mockingbird species has this property, though Darwin plausibly could have asked whether any mockingbird species inhabits just Charles Island as a prerequisite to seeking the identity of this species. Also *N. Trifasciatus* lives just on Charles island" is not an answer that Darwin would or could have offered to the question because to him the genus of South American mockingbirds was known under John Gould's description "*Orpheus*", not the expression "*Nesomimus*" introduced later into ornithological discourse. Thus "Snt&Cnt" cannot be Darwin's answer to his question, where "nt" stands for "*N. Trifasciatus*".

To express Darwin's answer, Sot&Cot, so that it entails the

desideratum to his question, (Ex)K<d>Sx&Cx, an expression must be conjoined to this answer that says "Darwin knows what *Orpheus Trifasciatus* is". Hintikka has chosen (Ex)K<d>(x=ot) to express this conjunct. Accordingly,

Sot&Cot&(Ex)K<d>(x=ot)

is a direct answer to Darwin's question, and correspondingly

Ps&As&(Ex)K<*i*>(x=s)

is a direct answer to person *i*'s question. As in the case of the question itself, *i*(Ex)K<d>(Sx&Cs), the answer Sot&Cot&(Ex)K<d>(x=ot) should be interpreted in the light of knowing what are the metascientific requirements for 'knowing species' in the contexts of inquiry that Darwin participated in. If the latter expression answers the question then (Ex)K<d>(x=ot) is true only if these requirements are met. We should recall that these requirements changed for Darwin just after he returned home from the *Beagle* voyage. Thus we will have to offer indices d, d', d'',... to indicate changes in the epistemic settings in which Darwin worked during his lifetime.

As to the distinction between minimal and adequate answers to these questions, we have already decided that the index *i* in the K<*i*> operator indicates a particular intensional setting in which person *i* asks the question. This setting includes a specific terminology and means of conceiving a subject matter as well as criteria for a suitable answer, in this case epistemological criteria for determining when that asked for in the question has been sufficiently identified. The K<*i*> operator in the direct answers has the same index, and accordingly the setting in which the (Ex)K<d>(x=ot) component of Darwin's answer is indicated to be the same as that in which his question was asked. This assures that the concepts and epistemic standards underlying his identification of species *O. Trifasciatus* are just the species concepts and metascientific desiderata presupposed in his asking the question "What mockingbirds are endemic to Charles Island?" The matching of indices is a requirement at least for direct answers to a question.

However, conceptual problems may call into question some essential component of the context of inquiry from which the question was initially conceived, believed valid and preferred. Also it may be discovered that factual questions have warrantedly credible answers that are incompatible or incoherent with the question's presupposition. Both conceptual innovations and factual discoveries can conflict with and require revision of extant metascientific standards, as examples discussed in earlier chapters have shown. In such cases *corrective answers* require revision in the questions asked or the epistemic background to the context of inquiry from which the question was initially asked. Here it would be inappropriate to require both question and answer to have the same index.

Darwin's Q2, "Are these specimens distinct species or distinct varieties?", and its subordinate, Q3, "Do the variations observable among the mockingbird specimens correspond to variations among acknowledged species?", as we have seen, were eventually given answers that entailed revisions of his initial Lyellian species concept. Q2, and Q3, when considered with his question Q4, "What kind of evidence would support the hypothesis that the neighboring populations are distinct species (one answer to Q2) if it were available?", suggest strongly that there is a conceptual component to Darwin's inquiries. His problem is one of interpreting the evidence that he gets in answering Q3, or in determining the taxonomical significance of that evidence. However, conceptual problems are not here to be construed as did the positivists and operationalists: The positivist construal is that solving a conceptual problem is to determine just by 'logical analysis' the empirical content of some concept C, such as Lyellian or post Lyellian species concepts. Darwin's Q4 was not answered by armchair logical analysis but rather by empirically querying scientific practice by means of Q3.

Darwin's conceptual problem in this case is his own problem in a sense, that is, it is like a pedagogical demand for an explanation where a student seeks an 'explanation' of some explanandum in terms already known to the scientific community. Answers to Q2 are sought by addressing the knowledge possessed by professional ornithologists. This kind of 'why-question' should be indexed so that it is ascribed to the individual, not to a scientific community. However, in another sense his question is also like the demand that a scientific community might make for an explanation, where the concepts and beliefs involved in answering the question are not in the possession of the community. The explanation of the mockingbird specimens that is sought in Q2, the variations among neighboring populations on the South American mainland and nearby islands and further answers to Q3 that might be forthcoming, are not common knowledge to the community of natural historians but are epistemic items newly introduced by Darwin. Hence Darwin's problem is in a sense a problem for the entire community and should be indexed over that community. On the other hand it does not pose a great challenge to knowledge possessed by that community and is, in fact, answered by appealing to standard species concepts among at least some members of that community. Thus, though Darwin's own epistemic circumstances, particularly his species concept, are modified by his eventual answer to Q2, we cannot say that those of the community are thus modified, at least beyond the addition of, albeit important, factual knowledge to the shared background ontology. Hence, at least in this case, common indexing of question and answer seems warranted, though the complexity of the circumstances of Darwin's inquiry cannot be adequately captured by this indexing.

One further important concept in questioning is a question's *presupposition*. Intuitively and notoriously the question "When did you resign from the Communist Party?" presupposes "You were once a member of the Communist Party". To call this sentence a 'presupposition' means that a person who never belonged to that party is under no obligation to answer this question. Thus the concept of presupposition is fundamental to any methodology for determining what questions should be asked under what circumstances. Also these intuitions suggest that the falsity of the presupposition is sufficient to defeat the question, that is, questions Q entail their presuppositions, Pres(Q). The entailment from Q to Pres(Q) is easily obtained from the fundamental epistemic principle (K1) in this way: The desideratum Core(Q) to Darwin's Q

$$(Ex)K\langle d\rangle(Sx\&Cx)$$

entails by (K1)

$$(Ex)Sx\&Cx$$

that is, that there are mockingbirds endemic to Charles Island. This proposition was mentioned above as a prerequisite to Darwin's inquiry, and thus both intuition and logical principles in QIE are satisfied.

It should also be noted that this expression for a question's presupposition has no epistemic operators. Thus the objectives of Darwin's inquiries, or the validity of his research programs, can be appraised in terms of their compatibility with what we take to be fact. As pointed out earlier, Darwin can have as one objective knowledge of, e.g., a species, which is belief supported by the kind of evidence that in his epistemic context would be certifiable as indicating or making it likely that the belief is true. These objectives fit what Darwin actually sought, viz. knowledge of actual genetic isolation of populations of animals and plants. Thus, contrary to Laudan's suggestions (1977), important scientists do pose problems in which truth, or correspondence in some sense to reality, is an objective. Problem based epistemologies can be realistic or pragmatic depending upon whether the basic problems are taken to be epistemic problems, like the questions discussed above, or pragmatic problems involving decisions, actions and the appraisal thereof in the light of values other than truth as correspondence to reality.

Since its expression has no epistemic operators, the presupposition (Ex)(Sx&Cx) is in no way committed to the beliefs and conceptions that constitute the epistemic setting for Darwin's inquiry. The expressions "Snt&Cnt" and "Sot&Cot" are both among its instantiations. Thus, however we conceive, describe and know mockingbirds, if there are none endemic to Charles Island, then Darwin's questions are false, even though he asked them, and believed himself correct in conceiving, expressing and presupposing what he did in asking his question. We are not advocating epistemic relativity here, according to which the validity of Darwin's questions can be assessed only within his epistemic setting. Aqvist's way of expressing presuppositions allows us to assess Darwin's questions in our own epistemic setting. Such assessment should be within the power of science and natural philosophy, even if it is not always good historical heuristics. Accordingly, using the language of QIE we can express with historiographical adequacy Darwin's epistemic aspirations and yet retain the power of expressing our appraisals of these aspirations. Darwin's epistemic aspirations need have no implications regarding the factual content of our actual world, except, of course, that this world includes Darwin and his aspirations. We can believe that Darwin aspired to know which species of *Orpheus* was endemic to Charles Island without ourselves believing that there are any endemic mockingbirds on that island and without necessarily conceiving and naming mockingbirds as Darwin did.

3.5. OTHER SCIENTIFIC QUESTIONS

One further question in deciding the usefulness of erotetic logic as a tool of scientific inquiry is whether we can obtain reasonable expressions in QIE for the variety of questions that might be asked in science within a field, whether that field be viewed synchronically (e.g. consisting of contemporaneous problems of epistemic determination, problems of explanation, pragmatic problems) or diachronically (e.g. the evolution or 'revolution' of some problem(s) of explanation as a field develops). No attempt will be made here to give an exhaustive answer to this question. (See Kleiner, 1985, for a more extended attempt to achieve adequacy in this sense.) However, we can review the formulation of some types of question that may arise at one time in a field and then we shall consider how diachronic shifts in questions might be handled in the language we are considering for erotetic logic.

We have already argued at length that epistemic questions, problems whose objective is to know something, such as the value of a universal constant or the value of some variable quantity such as mass, optical density, resistance, capacitance (problems of determination), can be formulated using the descriptive language of second order predicate logic and the modalities of epistemic and deontic logic. On the other hand, practical problems in science, that is problems concerning the choice of actions, seem to be not easily captured with the tools of deontic logic presently available. People, scientists included, have a variety of obligations that are not neatly nested into categorically obligatory objectives, hypothetically obligatory sub-objectives and various permissions and prohibitions. It is not clear, for example, how moral and epistemic obligations can be nested together: Are moral obligations fundamental and without condition and epistemic objectives in the service of moral objectives, as some pragmatists wish to claim, or can there be unconditional epistemic obligations, with or without some moral justification? It would seem that practical problems in science need to be better understood philosophically before one can seriously try to construct a logic that will do them justice. We shall continue focusing our attention on epistemic problems in science.

Darwin also asked certain historically significant questions about the Galapagos mockingbirds, particularly "Why are there species of mockingbird endemic to Charles Island?" and the equivalent for Albemarle Island, etc. From some knowledge of Darwin's actual inquiries, this expression is ambiguous in that it could convey at least two different kinds of question, both of which were items of interest to Darwin: (i) 'What is the biogeographical history of events that terminates in the presence of *Orpheus Trifasciatus* just on Charles Island?' Here, for Darwin, as opposed to some of his predecessors and contemporaries, such as Lyell, this history can include migrations, extinctions and speciations, but it also should not include creations or the birth of species from no reproductive ancestors. (ii) 'What is the causal mechanism by which reproductive ancestors of *O. Trifasciatus* who colonized Charles Island evolved into this species which lives just on this island, where this mechanism must be a genetic mechanism describing natural processes of variation and inheritance that occur throughout nature, in wild animals and plants as well as in humans and domestic animals and plants?'

Two points will be made immediately about these questions: First the disambiguation of the why-question, that is rewriting it so that it can explicitly express distinct questions actually asked by Darwin, is achieved by paraphrasing it as a what-question, a question with the same overall form as the two considered above (Cf. Hintikka, 1981). In case of (i), what is asked for is a history, i.e. a temporally ordered sequence of events that are described as terminating in the presence of a population identified as *O. Trifasciatus* and confined to Charles Island. This history includes migration and colonization events, extinction events, speciation processes, but excludes supernatural activity. Thus (i) (that is, its desideratum) can be expressed roughly as

$$(Ex)K<d>[Cx,ot \;\&\; Tx \;\&\; (Mx \vee Ex \vee Sx) \;\&\; \sim Cx]$$

where "Cx,ot" is "x terminates in the endemic presence of *O. Trifasciatus* on Charles Island", "Tx" is "x is part of a temporally ordered sequence of events that compose a phylogeny", "Mx" is "x migrates to Charles Island at some time", "Ex" is "x goes extinct at some time", "Sx" is "x appears at some time as a distinct species from some of its ancestors", and "Cx" is "x is created, i.e. x has no parent population but comes into existence at some time due to supernatural causes". "At some time" may be interpreted as referring to time intervals of various durations located by some date in phylogenic history. It is also another determinable whose determinations could be asked for in further phylogenetic inquiry. The phylogeny here includes branches (speciations in cladist taxonomy) and termini in temporal dimension (extinctions), viz. temporal radiation, as well as spreading and dividing in geographical space, viz. spatial radiation. This phylogeny is a history or a spatiotemporally unfolding individual, and the question asks for that segment of it that constitutes the ancestry of *O. Trifasciatus*. This description of a phylogeny is part of the Darwinian ontology underlying his question.

This formulation of one Darwinian why-question has the advantage of being a relatively simple expression of one historically important variant of the general biogeographical question "Why does species S reside in region R?" Lyell's variant of this question would include creations in the disjunction of possible answers, with migrations, extinctions and adaptations within species (the production of varieties and races) as well. In this respect Lyell's historical why-questions differ from those chosen by Darwin, and this distinguishing feature is an important component of the Darwinian revolution. Lyell's question, as distinct from Darwin's, has some ramifications for a background ontology, particularly in that Lyell permits individual species to have temporal beginnings without ancestors, whereas Darwin's ontology forbids such events. In Darwin's case, for all segments and termini of a phylogeny that are species, there are segments that are their reproductive ancestors. Hence there are differences in the connective structure of the phylogeny between Lyell's and Darwin's ontologies, though many spatiotemporal segments, that is many species that they would identify, are the same.

In (ii) Darwin is asking for some causal mechanism and a way of conceiving this method. However (ii) can also be split into two components: One component asks for concepts for describing productive relations among segments of his phylogeny and their ecological and geographical circumstances, in whose terms certain laws (Darwin's 'laws of life') can be formulated. The second component asks for knowledge of these laws. The laws sought connect by reproductive processes earlier and later segments of the phylogeny and connect by genetic and ecological relations attributes of these segments. Thus this one of Darwin's why-questions can be put in the following form:

$$(ER)K<d>[Ts \ \& \ T'ot \ \& \ Rs,ot \ \& \ G(R)]$$

where "Tx" means "x is a specific branch of mockingbird phylogeny located on the South American continent sometime in the past", "T′x" means that "x is a specific branch of mockingbird phylogeny that is presently on Charles Island", "s" designates a species that is an ancestor to *O. Trifasciatus*, R is a relation (Cf. Bromberger's 'W-relation', 1966) in the phylogeny to be found and "G(R)" is the second order property describing R as a genetic relation, that is a relation satisfying the genetic laws believed by Darwin.

By contrast, the creationist Lyell held that creations take place throughout phylogenetic history. Lyell would reject as false the presupposition

$$(ER)[Ts \;\&\; T'ot \;\&\; Rs,ot \;\&\; G(R)]$$

because, at least before around 1860, he did not believe that there were any genetic relations between species. The only genetic relations of this kind that could exist in Lyell's *weltanschauung* are relations within species, viz. relations within varieties and between varieties of one species. Lyell would thus require that any R for which G(R) holds must be reflexive when the relata are species. The corresponding Lyellian explanation for a speciation event would be to ask for the location of a speciation in geographical space and time, that is a spatiotemproal location for a parentless segment of phylogeny, a 'center of creation', and possibly for a 'purpose' that would explain the apparent adaptation of the species thus introduced into its surroundings. Lyell thus would reject Darwin's question as incompatible with his background ontology, that is, he would reject both of Darwin's why-questions on the grounds that the knowledge they propose as research objectives is excluded from his ontology and thus their presuppositions are false.

In sum, Lyell's and Darwin's why-questions are distinct by containing descriptive clauses in their desiderata and presuppositions that place different constraints on allowable answers. In the case of two kinds of 'why-question' appearing in biogeography, the one asking for a novel causal relation between a species and some antecedent condition and the second asking for a temporal sequence of events that terminates in some location and range for a species, the constraints take the form of assumed descriptions of what is being asked for.

One procedure for the appraisal of questions is to reject them on the grounds of the belief that their presuppositions are false. There can be several grounds for this belief: (i) The presupposition is incompatible with received scientific ontology, which presumably is supported by a tradition of successful scientific theories. (ii) The presupposition might be more directly challenged on empirical grounds: One might reject Darwin's question about Galapagos mockingbirds by showing that he mistakenly assumed that there were mockingbirds on those islands. It might be shown by a combination of conceptual and empirical arguments that the islands are and were in the early 19th century populated with thrushes and Darwin confused mockingbirds and thrushes. (Remember he was subject to similar confusions about the Galapagos finches.) Also we have been over some of the assumptions behind species identifications in the previous chapter from which it should be clear that Darwin's ornithological researches required some reconceptualization of species as well as empirical evidence about variations and distributions. (iii) In carrying out an investigation of some received question Q one finds an answer A that is incompatible with Q's presuppositions. In such a case one can find herself in what Sylvain Bromberger (1965, 1966) called a p-predicament: One believes that Q is valid, but all answers she can think of for Q are ruled out by empirically based beliefs or background ontology. In such a case one has a choice of modifying or giving up Q or of modifying background beliefs. An answer A in the first case is called a *corrective answer* to Q in which accepting A as an answer to Q one revises or gives up that portion of her ontology that is the presupposition of her question. Such an occurrence can be a form of serendipitous discovery, as discussed in an earlier chapter: In searching for an answer to Q one discovers an answer to Q' which proves to be as valuable or more valuable than what was sought in posing Q.

Corrective answers to a Q need not occur just in the radical transitions of science that separate Lyell and Darwin. As Darwin's concepts can evolve within one program of research, the attempt to answer the question of whether *O. Trifasciatus* was a distinct species of mockingbird, so also one's questions can change within such a program: At one point of early in the researches that Darwin reports in his 'Notebooks on Transmutation' (DeBeer, 1960, see Kleiner, 1988 for details and documentation of the story to follow) he expresses the belief that geographical isolation is the only condition that is causally relevant to preserving variant heritable characters. For Darwin the variant characters are the small heritable differences produced at birth that can be seen in a litter of puppies, some of whom are lankier than others, some have longer hair than others, and there may be marked variation in color among them. Excluded are 'sports' or large differences that produce variant characters that are mostly deleterious and thus not passed on to subsequent generations. Isolation prevents variant characters from blending back into the population, and a succession of isolation events can cause accumulation of variant characters that could produce sufficient differences for speciation. These beliefs about variation and his belief about the preservation of variants can be considered to be among Darwin's background beliefs B and included among the constraints of the question: "How was *O. Trifasciatus* produced as species unique to Charles I?" What differentiates Darwin's from creationist why-questions for speciations is that Darwinian constraints restrict speciation processes to genetic processes in reproduction, which include who has which parents or who gets to breed with whom, occurrences that are influenced by migrations and intermingling and separation of populations.

Darwin believes his question is correct, for he believes that *O. Trifasciatus* is a species of *Orpheus* appearing just on Charles Island (what some, such as Bromberger, 1966, call *the* presupposition to Darwin's question), and that there is a natural cause for this occurrence involving migration, colonization and reproduction on the part of ancestral mockingbirds. This second conjunction describes the causal relation Darwin seeks, and hence is also a presupposition to his inquiry since to justifiably seek a natural cause presupposes that such a cause exists. There seems to be every good reason to include this among the presuppositions to the questions that Darwin posed on this occasion, for these objectives and what they presuppose are essential to his inquiry, as can be seen if it is contrasted to a parallel inquiry that Lyell might have conducted in his pre-Darwinian years. For our purposes Lyell might be expected to behave as did Darwin on his visit to the Galapagos: He would seek the location of a 'center of creation' for each species he recognizes, and once such locations are specified his inquiry would stop. In short, Darwin was a paradigmatic Lyellian biogeographer through most of the *Beagle* voyage. In this research program inheritance, variation, colonization and migration are causally irrelevant to the placement of true species.

However, any answer Darwin can think of to his 1837 question of how such speciation took place, i.e. any proposal involving possible means of further geographical isolation of the first *Orpheus* colonizer and its descendants on Charles Island, is ruled out by his geographical background beliefs. No succession of isolation events within the confines of Charles Island, e.g. accidental colonization of new inaccessible regions the island, the separation of components of the population by the occurrence of new geographical or ecological barriers, such as desertification of some territory within their range or the opening of a new sea channel between parts of that Island that subsequently closed, is consistent with his beliefs about the geological and ecological history of the island. Yet such isolation within the boundaries of the island is needed to keep the new variants from the original colonizers from backcrossing into the parent population. In effect, once a colonizing variant mockingbird is isolated on Charles I. there is nothing plausible to Darwin at this point of his search that would cause it to continue to vary, and to preserve these variations, thereby adding to their differences from populations on neighboring islands. (This same problem seems to have arisen later in Darwin's thinking: See his 'Essay of 1844'.)

Darwin is thus in a p-perdicament: He *believes* that a question Q, "Why is *O. Trifasciatus* found only on Charles I.?" has a correct presupposition, which in Bromberger's analysis is just that *O. Trifasciatus* is endemic to Charles Island. However, in the light of Darwin's concepts, e.g. species concepts C, and his background beliefs about laws of variation of offspring and inheritance L, and his background beliefs about ecological and geological happenings in the Galapagos Islands, part of his B, he cannot think of acceptable answer to Q. The options available to him in such a case are either to revise one or more of L,C or B to accommodate answers to Q, or to replace Q with some Q' that will admit of answers coherent with L,C and B. If we follow Bromberger's theory of 'Why-questions', Darwin's option of revising Q amounts just to giving up the belief that *O. Trifasciatus* uniquely occupies Charles Island. This he might do under an indefinitely large number of possible scenarios, e.g. if he learned that this species can also be found on the South American continent, that the various mockingbird specimens were only varieties of one species, or that the birds he thought were endemic mockingbirds were actually endemic thrushes, or were thrushes with a global range. However, Darwin never has to give up his explanandum. Still it seems that he does revise the descriptions of the causal relation that he seeks, descriptions that on our analysis should be included in Q's presuppositions because they specify the objectives of the inquiry that Darwin undertakes in posing his why-question. These revisions occur as follows:

As his thoughts proceed, Darwin addresses his problem by revising background beliefs, in this case genetic laws L, about possible ways in which variation is produced and, consequently, preserved. He moves to include reproductive habits and heritable traits produced by habit among admissible mechanisms for producing variants. In making this revision he apparently hopes that he can establish that more variation might produce, by sheer force of numbers, variant characters that survive backcrossing by virtue of frequency of occurrence. This hope turns out to be futile, for the backcrossing problem continues to trouble him much later, in 1844, but still it is a plausible reason for a move very much in evidence in his notebooks (Cf. Provine, 1971). In his search for an explanation of speciation Darwin initially believed that variation was always adaptive and would produce an accumulation of adaptive changes among the birds on Charles I., thus causing them to depart further and further from parent populations. This view fit well with his 'Lamarckian' hypothesis that acquired habit and resultant structural adaptations, like a blacksmith's muscles, could become 'fixed' in inheritance. However, problems of coherence with background belief still remain: Darwin believed that there were few geographical differences among the Islands that were sufficient to account for adaptive differences. He recognized in the same observations that exemplified heritable variations, as in litters of puppies, plant seedlings, etc. that the variations that occur at birth, such as white color or longer hair, generally take place only in part of a litter and hence only in part of a population, and he came to realize in his inquiry that such variations are not always adaptations. Here we have a shift in Darwin's conception of what variations are beyond additions to what cause them: They are not necessarily inherently adaptive. As he knew earlier, because they occur only in part of the population they can be lost by blending back in backcrossing, and hence are not likely to accumulate even if they occur at a high rate among new offspring and are produced in the adaptations of adults to their behavioral needs. Longer hair or lighter color might be adaptations in colder climates but they offer no advantage and may even be deleterious in warmer climates. Thus background beliefs still rule out the likely accumulation of variations and the continued divergence of geographically isolated populations. They now also rule out inevitable and perfect adaptation of varying organisms to their surroundings, but this realization becomes an opening for Darwin to escape from his p-predicament. In October of 1838 Darwin's attention was drawn to another aspect of reproduction after re-reading Malthus: According to Malthus reproduction in humans and in nature nearly always involves the generation of offspring that exceed their parents in number, that is, *superfecundity* is pervasive in human and natural populations. The inevitable population pressure within superfecund species requires the premature death of most members of a breeding population. Thus advantaged variants may have less chance of being lost by blending with less advantaged variants because the latter are less likely to survive and reproduce. Thus Darwin discovers a new mechanism by which traits are sorted and preserved, and this answers his causal why-question coherently with his background beliefs and without further modifying the constraints on how variants can be produced.

Finally, our decision to express why-questions in science as special kinds of what-question provides a means of expressing any one of a number of items that might be requested by "Why p", e.g. "Why is *O. Trifasciatus* part of mockingbird phylogeny, one of whose branches is located presently on Charles Island (T′ot in the above formulation)?" Intuitively part of the question's presupposition is p, for we do not generally seek explanations of what we do not take to be fact. For a number of writers on this topic, e.g. Bromberger (1966), the explanandum is all that is presupposed by a question. However, this conflicts with out analysis of questions as proposals of objectives to be sought. Not only is an explanation for an alleged fact sought in posing a why-question, but also a certain kind of knowledge which, according to current metascience, counts as an explanation. There also may be exceptions to the idea that explanations presuppose the truth of their explananda: One variant of abduction can attribute plausibility to T if T explains something whose factuality we are not quite sure of.

T′ot is a conjunct in the presupposition we have formulated for Darwin's why-question, and thus our formulation of why-questions fits the intuition, which may not be entirely true, that only facts are appropriate objects of explanation. The qualification on the truth of the explanandum just mentioned could be grounds for rejecting the implications in erotetic logic for science as being unrealistically rigid for what is reasonably allowed in science. One might consider some p plausible just on the grounds that we have a ready explanation for it, and this plausibility could motivate an expensive effort to establish p on other grounds. Further qualifications will be drawn later about using erotetic logic as a source of methodological prescriptions for scientific inquiry.

The formulation of why-questions that some (e.g. Bunge, 1967) give

$$(\mathrm{Er})\mathrm{K}<i>(\mathrm{p\ because\ r})$$

is not very illuminating. "Because" is ambiguous and can be used to refer to logical or evidential relations. The two are not the same: I believe p because it follows deductively from r, which I believe; I believe p because reliable authority E told me so. 'Because' also can be used to refer to various causal or other natural relations: p happened because conditions r made it overwhelmingly probable, or p holds because it is a special case of symmetry principle S.

Presumably inquirers, such as Lyell and Darwin, can ask for rather specific causal relations or historical narratives in posing problems of explanation, and paraphrasing their questions as what questions provides the language of erotetic logic with the means of expressing what is distinctive in their questions. Furthermore, in part because of the obscurity of logical relations that might be implicit in 'because', there is no obvious deduction from this formulation of the why-question to its presupposition p, as there is in our formulation. Indeed we can infer

$$(Er)(p \text{ because } r)$$

from this formulation, and this is in accord with intuition. To ask "Why p?" presupposes that something either brought about p or is a reason for p. However, it is unclear how p can be inferred from this as a presupposition. For these reasons why-questions need to be reduced to what-questions in any erotetic logic which might be useful for science.

3.6. CONCLUSIONS

The Hintikka-Aqvist erotetic logic captures several important features of scientific questions. The epistemic part is appropriate for certain scientific questions, which are plausibly construed as expressing aspirations to knowledge, that is belief which is supported by evidence believed to be veridical. Also the epistemic expressions are intensional and incorporate indices referring to persons to whom a question is attributed. Generalizations of these indices can refer to communities sharing metascientific values and other background beliefs, and thus by designating distinct intellectual communities they can also be interpreted as designating distinct contexts of inquiry. Distinct indices can represent variations in the terminology, the conceptualizations and the epistemic values by which members of a scientific community at a given time set and express epistemic goals. Also indices indicating persons possessing or aspiring to knowledge occur in expressions for answers as well as questions in this version of erotetic logic. Thus like indices in the epistemic components of questions and answers can be used to indicate that the answer meets the epistemic conditions whose fulfillment is sought posing the question. However, in actual science answers need not meet such conditions but can establish different epistemic conditions requiring a modification of the context of inquiry, or a modification of the initial question that contributed heuristically to the generation of this answer. Darwinian background beliefs, such as the belief that all organisms are generated by established means of sexual or asexual reproduction, are incompatible with beliefs that species are created. Thus in Darwinian contexts of inquiry creationist why-questions are forbidden because their presuppositions permit the possible production of organisms without natural parentage, which would be incompatible with Darwinian background belief. Still it remains possible to pose corrective questions against received background belief, just as it is possible to discover corrective answers to received questions. The acceptance of corrective questions requires revisions in background beliefs, just as the acceptance of corrective answers requires revisions in the very questions that heuristically generated them initially. Since erotetic logic has virtually nothing to say about the nature of epistemic conditions or their valuation or methods for their valuation in particular scientific circumstances, it provides no insight into methods for resolving conflicts between epistemic conditions presupposed by questions and different conditions implicit in the knowledge of corrective answers to these questions. Parallel points can be made about justifying questions that are corrective of background beliefs.

The question-answer relation in this erotetic logic implies that knowledge of a question is knowledge of some or more attributes of possible answers. Such knowledge can accumulate or can be diluted in the process of an inquiry, processes which respectively can be construed as getting closer to or farther from an answer to the question. One important kind of heuristic is that which produces an accumulation of attributes to answers. Knowledge of an answer entails the desideratum of question, that is the fulfillment of the epistemic objective, or the descriptions of possible answers posed by the question, but not the converse. Thus unanswered questions can be known. Also answers can provide new conceptualizations as might be needed in identifying individuals that answer the question. Hence QIE can accommodate a broader range of questions, particularly questions demanding conceptualization, than those formulated in PIE, where questions are construed as explicit disjunctions of possible answers and thus presuppose a conceptualization of possible answers.

The deontic part of H-A logic captures the imperative component of a question, i.e. all questions are requests or demands for whatever constitutes their desiderata. The ordering of the priority of scientific objectives from urgency or utmost priority must be understood in terms of the appraisal of a question. The 'logic' or methodology of question appraisal is the topic of a chapter to follow.

The 'logic' of some corrective questions and answers can be represented in the H-A logic, viz. those questions with presuppositions that contradict background beliefs or those answers that contradict a question's presuppositions. However, as in sentences, to discover what constitutes contradictions in a scientific language requires knowledge of particular scientific concepts which cannot be derived from formal logic of any kind. Also actual science has further complexity that demonstrates the inadequacy of matching answers and questions by matching indices that designate individuals and communities in particular epistemic circumstances: Questions and answers that challenge various aspects of a context of inquiry behind the question can be complicated and cannot be adequately represented in H-A language. Darwin's question about taxonomical significance of Galapagos mockingbirds is a case in point: His own conceptualization of species is challenged by that in a related field, ornithology, but on the same occasion he introduces new biogeographical information to the ornithological community. Whether the question can be attributed just to him, in which case it would be a question posed just for his edification by knowledge already possessed by a scientific community, or whether it represents a defect in the prevailing epistemic context of his field or related fields, turns out to be a complex matter that cannot be expressed by indexing the question over a scientific community or by matching the question-answer indices provided in the language of H-A logic. The explanandum of Darwin's question, namely the distribution of South American mockingbirds, is not common knowledge among biogeographers and ornithologists. However, the concept he seeks, which turn out to be a species concept in the context of inquiry of the ornithological community in the spring of 1837, is common knowledge in that community. Hence in one aspect this question is directed just at Darwin's edification, the provision of a concept adequate to account for mockingbird distribution. On the other hand the answers to Darwin's question edify both biographers and ornithologists, and surely others as well. Darwin's biogeographical data when conceptualized with concepts common in ornithological practice strongly support the conclusion that species transmutation accompanies colonization.

Thus H-A erotetic logic gives little insight into actual contexts of scientific inquiry beyond providing abstract conditions for permitting and forbidding questions and for matching epistemic conditions for questions and answers. The complexities of attributing questions to individuals or communities, distinguishing those with answers available from knowledge established in the community from situations in which the question indicates deficiencies in community knowledge cannot be found in the abstract study of erotetic logic. Rather they must be discovered by analysis of actual historical circumstances and can vary from one to another of these circumstances.

Correspondingly erotetic logic tells us nothing about what it is to know answers to questions, though in actuality such conditions may be implicit in the knowledge of a question. Conditions for knowledge such as the identification of concepts or of species must be brought out by philosophical analysis of various concepts, such as species, number, individual classmate, politician, philosopher.

CHAPTER FOUR

A METHODOLOGY FOR PROBLEM CHOICE

4.1. INTRODUCTION

In this chapter the kinds of argument that can be offered in defense of the choice of a scientific problem will be considered. What problems to choose is a question facing every scientist who makes decisions in or about research programs, yet generally principles guiding this choice are neglected by philosophers.

Kuhn (1970) and Laudan (1977) have suggested that this kind of decision is made by joining a scientific community whose conduct is guided by an established paradigm, or whose goals and procedures are prescribed by an established research tradition. Their suggestion is that all problems are essential components of paradigms or research traditions. However, this suggestion begs the question regarding reasons for choosing questions, for if the decision to pursue a problem is a decision to join a research tradition, then we can ask for appropriate reasons for this last choice. Also the boundaries of research traditions are vague in the sense that there are no strict conditions regarding what counts as a change within as opposed to a change of a tradition. It is thus not clear what kind of problemshift, or change of the weighting of a problem constitutes a paradigm change or a revolutionary termination of one and initiation of another tradition. Problems of incommensurability suggest that one should avoid construing identity conditions for research traditions so that small changes in content count as revolutionary (Feyerabend, 1962, Shapere, 1966). This argument can be extended to recommend that research traditions should be defined to allow that certain weightings of problems and perhaps even certain kinds of problemshift can occur within a single tradition and can be defended by appeal to methodological and substantive beliefs that are part of that tradition or possibly shared with other traditions.

Furthermore, the choice of problems not sanctioned by an established paradigm or tradition must be considered. If revolutionary science can be embarked upon for good reasons, there may be good grounds for choosing problems that are incompatible with one or more of the several components of an established tradition described earlier: cosmological B, metascientific M, conceptual C, nomological L, and empirical R,E. Kuhn (1970) and Lakatos (1970) suggest that 'problemshifts' are occasioned by systematic failure to generate empirically adequate theoretical models T,T′,..., where in revolutionary science blame for failure is placed on all of the elements B,C,L,R,E of the established belief structure of practicing scientists in a given discipline. However, this kind of holism in scientific revolutions has been discredited (Laudan, 1984). It can be argued as follows that this holism unrealistically rules out the possibility of revolutionary research programs:

Changes in M, e.g. metascientific standards imported from other disciplines, may raise conceptual problems in the complex B,C,L,R underlying an established research program. Thus Lyell and Darwin respectively import *vera causa* requirements from newtonian physics. These requirements demand that only established or true causes be attributed as hidden causes of observable effects. Newton's *vera causa* principle stipulates that true causes be those that can be observed within the reach of experiments (1962, Bk III, pp. 398ff). Lyell thus departs from the family of traditions in which catastrophic 'revolutions' are permitted in geological history because they, or their causes, are not the kind of occurrences presently observable. One is thus faced with the epistemological question of how the latter causes can ever be known. Lyell thus argues from an epistemic objective, that knowable or known causes be cited to explain past geological events, to a methodological conclusion, that all geological explanations be in terms of causes presently observable.

Darwin also departs from the family of research traditions that allow non-reproductive creationist 'introductions' of new species, a kind of explanation initially endorsed by Lyell. In the place of creation events and centers of creation Darwin introduces candidates for genetic mechanisms that produce and accumulate variations, and the concepts used in describing these mechanisms are borrowed from common knowledge of the production domestic plant and animal varieties (see Kleiner, 1988, for details).

In effect, Lyell and Darwin choose conceptual problems incompatible with the research traditions respectively in geology and natural history which were widely subscribed to in the 19th century. A Lyellian "What caused this mountain range?" or "Why do the Alps exist?" contains implicit limitations upon possible answers that exclude geological forces not acting presently or with present intensities. Similarly, a Darwinian "Why are there distinct mockingbird species on Charles and Albemarle Islands?" implicitly excludes creation events as possible answers. These problems arise in their respective research traditions from a perception on the part of Lyell and Darwin of a lack of explanatory power and epistemic accessibility in certain of the belief structures B,C,L or models $T, T', \ldots$ in the preceding traditions of catastrophist geology and creationist biogeography. These deficiencies are assessed in the light of certain metascientific beliefs M largely based upon metascientific attributes of the enormously successful newtonian physics.

Both Lyell and Darwin consequently introduce significant revisions in the metaphysical background, the conceptual and nomological apparatus, and the evidential relations of respectively geology and biogeography. In Lyell's case laws of elevation, deposition and erosion are the same throughout geological time, 'primary' formations are not to be construed as remnants of the Earth's initial state, and evidence of past formations and events are linked to the present by means of these laws. Geological events leave traces, e.g. eroded sedimentary layers above volcanic rock, that are interpreted in terms of presently acting forces of volcanic and sedimentary deposition and erosion by aqueous and atmospheric forces. In Darwin's case, laws governing reproduction were extended from present observations of domesticated species and humans to wild species past and present, laws of superfecundity, variation and selection were introduced, and speciation was conceived in terms of geographical isolation, evolution and genetic isolation. As in Lyell's case, Darwin interpreted the fossil record in terms of presently tested geological causes and processes.

Changes in evidential relations R due to the development of successful theoretical models T,T′,... in other disciplines can also motivate revolutionary changes. Thus discovery of Roentgen rays (X-rays), their subsumption under Maxwellian theory of radiation and their application to simple repetitive crystals opened opportunities for studying more complex, less regular crystals in biochemical substances, such as various natural and artificial polymers, proteins, and nucleic acids (Olby, 1974). Other evidential relations and experimental techniques developed in microgenetics, including the use of radioactive isotopes as markers, the growth of mutant strains of mold and bacteria in variously deficient nutritive media, the experimental use of enzymes with specific activities for breaking up large molecules in specific ways, and the use of infective viruses, bacteriophages of selected strains to modify bacterial genomes. These methods made possible the experimental study of (a) large organic and artificial molecules' structural characteristics, (b) biochemical pathways such as heritable deficiencies in enzyme production and their effects on the vitality of microorganisms, (c) the function of DNA as opposed to protein as the genetic material, (d) the fractionation of proteins and nucleic acids and the identification of their components and their quantitative assay for various species by electrophoresis and chromatography. These new experimental methods, in some cases consisting of new evidential relations derived from advances in other fields such as physics and in other cases based on the application of established genetic and chemical procedures to microorganisms, made possible the importation of biochemical concepts C into the domain of genetics. This process included some revision of traditional biochemical concepts C to C′, e.g. concerning the possibility of large macromolecules, and genetic concepts Cg to Cg′, e.g. the division of genetic units into distinct units of mutation, function and recombination. According to one philosopher (Kitcher, 1982) these steps had as much conceptual significance to constitute the elimination of the traditional gene concept. Within this newly constituted genetic enterprise are sought families of interfield models T,T′,T″,... (Schaffner, 1980, Maull, 1977), including structural determinations of genetic materials and determinations of biochemical pathways that will account for the expression of heritable characteristics in microorganisms and, hopefully and ultimately, in all animals and plants.

These conceptual developments gave specificity to the questions "What is the gene?" and "How do genes act in producing distinctive observable characters of the organism?" The first question can be further specified: "What is the structure of the chemical substance, DNA, that carries genetic information?" and "What atomic or molecular structure explains genetic specificity in DNA?" The second can also be given further specificity: "What are the chemical pathways by which the genetically specific characters of DNA are expressed in the presence of enzyme X or structural protein Y in the metabolism and tissues of organism Z?" Adding such specificity can be construed formally as adding specific descriptions in the questions "What defines and distinguishes x as a biological organ?", "What is the causal relation R by which Px produces Qy?" to produce respectively "What P&Q define and distinguish x?" and "What is the causal relation R by which Px produces Qy and C(R)?". Here C(R) characterizes the relation R as a kind of causal process, a biochemical process. In addition to permitting these more specific questions of molecular genetics, the above developments enhanced their worthiness for the time and attention of geneticists by providing available and promising means for producing their answers. (See the discussion of the computer program PI below in which the initial and goal states of questions activate concepts relevant in the investigation.) Thus the biochemical revolution in genetics is methodological and experimental as well as conceptual, and it includes a significant change in the explanatory aspirations of geneticists, as can be expressed in questions like those just mentioned.

This wide-reaching revolution in genetics shares many features of the Kuhnian conception of a scientific revolution, but there are significant departures from that conception as well. The biochemical revolution began not from anomalies or persistent empirical mismatches in traditional Mendelian and cytological genetics, but rather was motivated in part by the metaphysical background belief that life processes are fundamentally chemical, and thus genetic processes ideally should be described and explained in the kind of causal detail apparent in other biochemical models. Secondly, the development of experimental means for achieving this ideal was an important factor in motivating its pursuit. Demonstrating that knowledge of entities and processes in chemical genetics is possible vindicates as an objective the chemical explanation of genetic processes, thus showing that objectives sanctioned by a materialistic biological metaphysics are not utopian (Laudan, 1988).

These examples of non-Kuhnian and quasi-Kuhnian revolutions suggest that there can be good reasons for choosing problems or research objectives as well as means of pursuing these objectives in circumstances in which neither the ends or means are sanctioned by established scientific paradigms or research traditions. These objectives can be made more or less specific by adding or removing descriptions in the 'What-questions' that can express the objectives. Good reasons for the objectives are good reasons for choosing or revising and for elevating in importance questions that express the objectives. It is thus that 'extraordinary' research can have a rationality or 'logic'. Extraordinary research is seldom, if ever, the undirected random exploration of science in crisis envisioned by Kuhn or the arbitrary and sometimes reversible shift in interest and other decisions envisioned by Lakatos and Feyerabend.

4.2. EPISTEMIC OBLIGATIONS

In this section postulates for PIE-QIE will be chosen and interpreted in the light of their scientific utility, particularly as they appear to bear upon problem choice.

One obvious contribution of erotetic logic to guiding and understanding scientific reasoning concerns the presuppositions that can be inferred from a question, as

$$(Ex)Px$$

may be inferred from the 'what-question'

$$!(Ex)K<i>Px$$

If credible developments occur in the ontological, conceptual or nomological foundations of a discipline that are incompatible with presuppositions of problems previously studied, then traditional programs in the discipline addressed to certain problems can be deductively falsified. But is falsification a sufficient methodology for the choice of problems? This question might be considered in the light of the discussion of the Popperian claim that theory falsification is a sufficient methodology for theory choice. In both cases falsificationism provides no methodological account for the constraining effects of affirmed belief. It gives no rationale for guiding or requiring actions in accord with belief, where to believe a proposition p is to take p as true. Presumably the novel metaphysical and conceptual components B',C',L' conflicting with the foundations of a prior research traditions must have established credibility to falsify the latter foundations, just as one must establish the objectivity and repeatability of observational claims to falsify theories in Popperian methodology. Thus, as in the case of theory choice, falsificationism is insufficient for problem choice, and one must look for criteria for appraisal of metaphysical, conceptual and nomological propositions beyond the logical relations available in erotetic logic, just as one must look beyond deductive logic for criteria of appraisal in the so-called logic of justification for empirical science.

There are plausibly other significant logical relations that bear on problem choice, particularly the relation of entailment between questions. Presumably the preference of a problem entails the preference of at least those problems it is known to imply. The problems implied by a given problem can be construed as its subproblems. As can be shown in any adequate formulation of erotetic logic, P' is a subproblem of P just if P's solution is a necessary condition for the solution of P. From this it can be inferred that if P is valued, and P' is known to be a component or subproblem of P, so also must P' be valued. We shall attempt to appraise the methodological ramifications of this relation below.

There may well be further extralogical considerations that must be brought to bear in the choice of problems. These can be identified in part where logical relations are absent from erotetic logic. Deductive logic itself can be appealed to in demonstrating its own insufficiency for certain kinds of putative scientific inference. Hypotheses cannot be validly deduced just from the verification of their deductive consequences. More must be known in the way of extralogical evidential relations for the verification and confirmation of hypotheses. Similar demonstrations of insufficiency may be available in erotetic logic in regard to the rational support of problems. As in the case of putative 'inductive logics', whether this insufficiency should be attributed to methodologically incorrect science or to inadequate methodologies is an item for philosophical discussion and adjudication.

To specify what can and cannot be deduced from problems, answers and presuppositions, as sanctioned in available formulations of PIE-QIE, the following postulates and rules concern epistemic states and obligations regarding propositions (Aqvist, 1975, p. 27). These principles govern both the logic of unquantified questions, propositional imperative epistemic logic (PIE), and the logic of quantified questions. The deontic part will be separated from the epistemic part by prefixes D and K in the enumeration of these principles. Also, at this stage of the exposition '⊢' means that what follows it is a theorem (an unproved postulate or a proven assertion) in PIE-QIE.

$$\text{(D1)}\ \ \vdash\ !\mathrm{p} \rightarrow \mathrm{ip}$$

This is the intuitive principle that anything required must be permitted. The converse, anything that is permitted must be required, does not hold.

$$\text{(D2)}\ \ \vdash\ !(\mathrm{p} \rightarrow \mathrm{q}) \rightarrow (!\mathrm{p} \rightarrow !\mathrm{q})$$

If it is obligatory that p not occur without q, then obligations to bring about p imply obligations to bring about q. For example, if it is obligatory that I not drive without a license, then if I am obliged to drive, I am obliged to have a license. Again, this principle accords with intuitions.

$$\text{(D3)}\ \ \vdash \mathrm{ip} \leftrightarrow {\sim}!{\sim}\mathrm{p}$$

Here p is permitted if and only if ∼p is not obligatory. Smoking is permitted in this restaurant if and only if it is not the case that one must not smoke there.

Thus far, the deontic postulates seem to cohere with defensible intuitions. This much cannot be said for the epistemic postulates, however. (The following discussion is largely derived from Aqvist, 1975, pp. 27–37.)

$$\text{(K1)} \quad \vdash K<i>p \rightarrow p$$

This postulate is intuitive and seemingly defensible. If knowledge of p is ascribed to person *i*, then p must be the case. If Darwin knows that gemmules determine sex, then gemmules determine sex. But gemmules do not determine sex. Hence Darwin does not (and never did) actually know that gemmules determine sex, though he might have thought, believed or claimed so.

$$\text{(K2)} \quad \vdash K<i>(p \rightarrow q) \rightarrow (K<i>p \rightarrow K<i>q)$$

(K2) initially seems plausible: If the conditional p→q is known and p is known to *i*, then q should be known to *i*. However, suppose *i* is a logical ignoramus with no knowledge of the principle of detachment. Presumably a person could know that heavy rain always wets streets and that heavy rain is occurring without drawing the conclusion obvious to so many of us.

$$\text{(K3)} \quad \vdash P<i>p \leftrightarrow \sim K<i>\sim p$$

"$P<i>p$" is read "For all *i* knows, p" or "p is compatible with what *i* knows". The first half of this biconditional

$$\text{(K3.1)} \quad \vdash P<i>p \rightarrow \sim K<i>\sim p$$

says that if p is compatible with what *i* knows then *i* does not know that p is false. This principle seems unobjectionable. However,

$$\text{(K3.2)} \quad \vdash \sim K<i>\sim p \rightarrow P<i>p$$

is not defensible. If *i* doesn't know that p is false, it may be because of his or her ignorance of consequences of other items of his or her knowledge. p could thus be incompatible with these consequences without *i* knowing that p is false. Dry streets are thus incompatible with the logical ignoramus' knowledge that heavy rains wet streets and that heavy rains are occurring, though he or she may not know this incompatibility.

$$(\mathrm{K4})\ \vdash \mathrm{K}\langle i\rangle \mathrm{p} \rightarrow \mathrm{K}\langle i\rangle \mathrm{K}\langle i\rangle \mathrm{p}$$

That is, if i knows p, then he or she knows that he or she knows p. This principle is debatable if not totally implausible to the thoughtful epistemologist. On the one hand, it does seem plausible to deny knowledge of p to a person who cannot provide normally acceptable evidence for p. Lucky guessers are not knowers. Yet it seems less plausible to say that knowing p implies knowing that p is known in the sense that knowing that this normally acceptable evidence is epistemologically certified. If this claim were allowed it would seem to have the consequence that only epistemologists can know, e.g. that it is now raining heavily outside. However, not all epistemologists agree that such knowledge claims can be made or upon the epistemological certification of normally acceptable visual evidence. Perhaps K4 might be replaced by

$$(\mathrm{K4+})\ \vdash \mathrm{K}\langle i\rangle \mathrm{p} \rightarrow !\mathrm{K}\langle i\rangle \mathrm{K}\langle i\rangle \mathrm{p}$$

according to which everyone who knows p should do some meta-science, particularly seek epistemological certification of his or her knowledge claims. These principles do not influence any of the arguments to be discussed below, however, where a rather different approach will be taken to epistemic evidence.

The following rule

$$(\mathrm{RK1})\ \text{If} \vdash \mathrm{p} \text{ then} \vdash \mathrm{K}\langle i\rangle \mathrm{p}$$

suggests that i knows all theorems of PIE-QIE, which is absurd. i need not be a logical ignoramus to fall short on this matter. All contingent beings, in the Aristotelian sense of "contingency", will, because of his or her finite lifespans, never be able to prove all of the infinite number of theorems in PIE-QIE.

Hintikka has suggested that $\mathrm{K}\langle i\rangle \mathrm{p}$ represent epistemic dispositions of an ideally rational person who is disposed to believe anything provided an acceptable proof is given for it. Thus, as long as i is not an incurable logical ignoramus he or she need not live for an infinitely long time to be disposed to believe any theorem proved in PIE-QIE.

Aqvist offers a weaker variant of these postulates and rules for what he calls PIE-QIE+. These accommodate what Hintikka has in mind and take advantage of the imperative component of erotetic logic in a way that is interesting for those pursuing the methodology of questions and of science.

$$(\text{K2+}) \vdash ![K\langle i\rangle(p \rightarrow q) \rightarrow (K\langle i\rangle p \rightarrow K\langle i\rangle q)]$$

(K2+) does not exclude the possibility that there are logical ignoramuses, but plausibly implies that in any epistemically ideal world there should be no such individuals. This normative assertion represents plausible aspirations of competent philosophers and competent philosophical pedagogues in present day epistemic circumstances, viz. that all deductive consequences of one's knowledge be known. However such a demand is not capable of fulfillment because there are often an infinite number of such consequences. Hence for heuristic reasons, viz. reasons concerning the practicability of certain tasks, there must be some further ordering of the importance of consequences whose knowledge is to be sought. In other epistemic circumstances indicated by various values of the index i the obligations might be different in detail, including knowledge of certain deductions $K\langle i\rangle(p \rightarrow q)$ and initial conditions $K\langle i\rangle p$. For example we would not demand that Aristotle know deductions sanctioned in Newton's theory of fluxions or in PIE-QIE. Also the obligation to know logical consequences of propositions in a given epistemic setting is one of possibly several general epistemic imperatives. However, because epistemic obligations are relativized to epistemic settings, they cannot, at least as obligations to know particular deductions and propositions, be considered categorical epistemic imperatives.

$$(\text{K3.2+}) \vdash !(\sim K\langle i\rangle\sim p \rightarrow P\langle i\rangle p)$$

(K3.2+) implies that all i should know what is excluded by the consequences of his or her knowledge in any epistemically ideal world. Again, this principle plausibly represents appropriate philosophical aspirations. It might also be a general epistemic imperative for scientists: Always explore the consequences of what you know as a means to determining what is compatible with your knowledge. In turn this latter determination may provide presuppositions to further questions that *may* be asked, questions whose presuppositions are compatible with current knowledge or knowledge in a given epistemic context. However, this principle still does not present obligations sufficient for the conduct of science: A number of philosophers have held that in a 'context of discovery' one can explore consequences of, and possibilities compatible with, hypotheses, theories, etc. which are 'entertained' as working hypotheses rather than believed, much less known. This is not to say that the context of discovery is exclusively the exploitation of working hypotheses, but surely the exploitation of possibilities incompatible with current background knowledge is heuristically valuable. Hence (K3.2+) can be considered too restrictive if it is interpreted as limiting what is permitted as possibilities to be explored to that which is compatible with received knowledge in a given context.

$$\text{(RK1+)}\quad \text{If } \vdash p \text{ then } \vdash !K<i>p$$

According to (RK1+), any theorem in PIE-QIE should be known in any epistemic context. This principle is too strong in that it presumes that PIE-QIE must be taken as infallible, or as knowledge in all epistemic settings. It is objectionable as a categorical epistemic obligation that is not relativized to a particular epistemic setting.

One further problem arises when the deontic and epistemic components of PIE-QIE are combined. One can obtain these as theorems in PIE-QIE+ (Cf. Aqvist, 1975, p. 3)

$$\text{(T1+)}\quad !Kp \rightarrow !p$$
$$\text{(T2+)}\quad !Kp \rightarrow ip$$

The first of these is obtained from (K1),

$$\vdash K<i>p \rightarrow p$$

from which, by (R1+) can be obtained

$$\vdash !(K<i>p \rightarrow p)$$

and which, when Kp and p are substituted into (D2) gives

$$!(K<i>p \rightarrow p) \rightarrow (!K<i>p \rightarrow !p)$$

and detaching the consequent, yields (T1+). (T2+) follows from (T1+) and (D1).

(T1+) and (T2+) imply, respectively, that if we are obliged to know that apartheid exists in South Africa then apartheid both ought to exist and may exist, a consequence that is contrary to robust moral intuitions. Aqvist's way out of this difficulty is to make a move that effectively restricts ! and i to epistemic obligations and permissions. By adding this postulate (D4): If p is an ordinary (non-epistemic) proposition, then

$$\vdash \mathrm{p} \rightarrow \, !\mathrm{p}$$

(T1+) and (T2+) then become equivalent to

$$(\mathrm{T3+}) \quad \vdash \, !\mathrm{K}<i>\mathrm{p} \rightarrow \mathrm{p}$$

which is unobjectionable and important as the basis for inferring presuppositions from obligatory problems by the simple operation of removing the deontic and epistemic prefixes "!" and "$\mathrm{K}<i>$". Aqvist's move effectively separates erotetic logic from other pragmatic or deontic logics where there are expressions and principles prescribing or permitting the actualization of non-epistemic proposition, i.e. principles prescribing and permitting non-epistemic actions, such as creating genetically engineered organisms or introducing them into the environment. Such actions are clearly part of the scientific enterprise and are even part of the epistemic enterprise if they are essential to conducting epistemically relevant experiments. Aqivst suggests that they can be accommodated by grafting other kinds of deontic operators onto PIE-QIE.

One other alternative would be to accommodate T1+ and T2+ by holding that there is no categorical obligation to know the existence of apartheid in South Africa. Thus we might argue from K2+ and RK1+ that all instances of logical conditionals must be brought about since we must know the implications of our beliefs and actions. Evils that are the consequents of counterfactual conditionals need not disturb us because they are only entertained in thought and do not exist. However, it is contrary to plausible intuitions that we are not categorically obligated to know the great atrocities of the world, especially if we have categorical obligations to prevent their recurrence or to mitigate those of their effects within our sphere of action.

We shall now turn to the scientific import of the PIE+ theorem

$$(\mathrm{T3+}) \quad \vdash \, !\mathrm{K}<i>\mathrm{p} \rightarrow \mathrm{p}$$

This theorem has a counterpart in QIE+

$$(T3.1+)\ \vdash\ !(Ex)K<i>Px \rightarrow (Ex)Px$$

which is readily inferred from (T3+). These theorems are important for inferring presuppositions $pv{\sim}p$ and $(Ex)Px$ from the respective questions

$$!(K<i>pvK<i>{\sim}p)\ \text{(Is p true?)}$$

and

$$(Ex)!K<i>Px\ \text{(What has P?)}.$$

These inferences are justified by the semantic stipulation that epistemic ideal worlds differ from the actual world only in respect to being distinguished by unactualized epistemic states and that questions hold in all ideal epistemic alternative worlds. Since each ideal epistemic alternative world differs from the actual world in terms of containing at least one non-actual epistemic state, the converse principle

$$(CT3+)\ p \rightarrow\ !K<i>p$$

and the corresponding QIE principle fails to hold. That is, problems cannot be deduced from their presuppositions. The contingent truth of a proposition p does not entail that p is known or that p should be known. If p is consistent p will hold in one possible world, but this world need not be one of the ideal epistemic worlds in which $K<i>p$ holds or in which $K<i>{\sim}p$ holds.

It is on the basis of (T3+) that Darwin can be said to be justified in rejecting a Lyellian question. It should be remembered from what was said earlier that Lyell's problem of speciation presupposes that species can be introduced without reproductive ancestry. In denying this claim, Darwin rejects Lyell's presupposition and, by (T3+), putatively falsifies Lyell's question. If Darwin can then establish as true his belief that all organisms in any phylogenetic line have reproductive ancestors, whether or not of the same species, he has then falsified Lyell's problem and the Lyellian program for explaining speciation.

However, the rejection of one set of problems is not itself sufficient to establish another set, as is indicated by the failure of (CT3+). Thus Darwin's grounds for rejection of Lyell's creationist program for explaining speciation and adaptation, and his rejection of problems that permit supernatural creation as solutions are alone not sufficient to establish his own research program. They do not suffice to justify the search for a genetic mechanism sufficient to produce adaptations and speciations. Even if he could establish beyond reasonable doubt that there is a genetic mechanism that can produce speciations with adaptations, problems of describing this mechanism with some specific degree of specificity, or of finding uniform features or governing laws do not necessarily follow. How much specificity should be required in further descriptions to be sought? Why should concepts manifesting uniformity in time or uniformity across species be sought? Erotetic logic does not answer these questions. Darwin's own argument is that natural history and biogeography should emulate physics in its search for 'simple' and 'sublime' laws. These objectives are based upon metascientific beliefs M held by Darwin and drawn from philosophers of his time, e.g. Herschel and Whewell, and are not in any way part of erotetic logic, particularly PIE-QIE+.

At this point it can be concluded that deontic logic is far from accommodating the various epistemic and practical obligations of the scientist. It does seem plausible that epistemic obligations should be distinguished from pragmatic and moral obligations, but nevertheless there may be important relations among them that govern the practice of science. For example, epistemic obligations can transfer in PIE-QIE to logical subproblems P′ of a problem P: If knowing p is logically necessary for knowing q then whatever epistemic obligation we have for knowing q we also have for knowing p. However, suppose that doing r, that is bringing about a state of affairs described by r is necessary for knowing p, as plausibly could be the case if some experiment must be performed in order to obtain knowledge of some p. In such cases there are connections between practical and epistemic obligations. Also practical and epistemic obligations may be linked to moral obligations if the occurrence of r has consequences that threaten certain harms to persons, the environment or possibly non-human organisms. In sum, an adequate account of the interlocking epistemic, pragmatic and moral permissions and obligations that constrain the conduct of science is yet beyond the scope of current deontic logics. Furthermore erotetic logic plays a role in the falsification of problems by allowing inferences via (T3+) from presuppositions believed false to false problems. However, there is no logical inference from a background belief taken to be true to a problem. Historical examples show that such inferences must be based upon extralogical metascientific beliefs, beliefs about epistemically relevant and other characteristics of good scientific theories that are established by historical precedent.

4.3. DEDUCTIONS, SUBPROBLEMS AND EVIDENCE

As suggested in discussions in earlier chapters, there are two general subproblems to epistemic problems, viz. problems of knowing p: (i) to determine which propositions q can count relevant as evidence for p and (ii) to produce q or conditions in which q can be reliably affirmed, such as conditions in which perceptions or observations that q can take place or conditions in which q can be reliably inferred from perceptions or from other beliefs. In this section we shall explore what illumination PIE-QIE can provide regarding this and other kinds of subproblem. If P′ is necessary for P's solution, whatever value is attached to P must also be attached to P′. Such a transfer of value is analogous to the relation of entailment between two propositions p and q: If p entails q, then whatever degree of belief is appropriate for p is also so for q. In the case of propositions a similar relation holds for material implication, where $p \rightarrow q$ is taken to be true but is not a logical theorem.

Suppose that $p \rightarrow q$ is a theorem in PIE-QIE+

$$(1)\ \vdash p \rightarrow q$$

then by (RK1+)

$$(2)\ \vdash !K<i>(p \rightarrow q)$$

Applying (D2) to (K2+) gives

$$(3)\ \vdash !K<i>(p \rightarrow q) \rightarrow (!K<i>p \rightarrow !K<i>q)$$

from which we can infer

$$(4)\ \vdash !K<i>p \rightarrow !K<i>q$$

Thus, if we are given a demand that p be known

$$(5)\quad !K<i>p$$

it follows deductively that if q is among p's logical consequences (by (1)), then

$$(6)\quad !K<i>q$$

Suppose $K<i>p$ holds in any ideal world, u^+, as required by (5). By (1) $p \rightarrow q$ holds in all u^+. (RK1+) tells us that all theorems should be known, that is $K<i>(p \rightarrow q)$ is in all ideal alternatives to the actual world u^+.

(K2+) tells us that $K<i>(p \rightarrow q) \rightarrow (K<i>p \rightarrow K<i>q)$ also holds in all such ideal worlds. Since $K<i>(p \rightarrow q)$ holds in all u^+, so does $K<i>p \rightarrow K<i>q$. Then, if $K<i>p$ is in all u^+, so is $K<i>q$. Thus all logical consequences of items whose knowledge is demanded are themselves to be demanded.

An inspection of this semantic argument shows that $p \rightarrow q$ need not be a theorem in PIE-QIE+ for the inference from (5) to (6) to be sanctioned. Suppose just $!K(p \rightarrow q)$, where $p \rightarrow q$ need not be a logical truth of any kind, but might be just some condition where p does not occur without q. Then $K<i>(p \rightarrow q)$ is in all u^+ then if $K<i>p$ is in all u^+, so is $K<i>q$. Thus any demand for knowledge of some kind of implication, whether material or nomological, permits transfer of problematicity from its antecedent to consequent. This condition $!K<i>(p \rightarrow q)$ cannot be further weakened, however. If $K<i>(p \rightarrow q)$ is not demanded, but is only permitted by what is known, then the epistemic alternative worlds in which it holds need not be those in which $K<i>p$ holds, and thus it cannot be inferred that $K<i>q$ holds in the same ideal alternative in which $K<i>p$ holds.

It is too much heuristically to assume that science demands that we seek knowledge of all conditional relations. Such a demand would overwhelm any available epistemic resources. The above considerations show that no such demand is entailed in PIE-QIE. However it is more heuristically plausible to assume that this obligation extends to nomological conditionals, viz. limited numbers of laws in each of the various domains of scientific inquiry. Nomological conditionals, which will be expressed by using '$\twoheadrightarrow$' as the connective, place constraints on all possible sets of models u of a given theory. If such a theory is paradigmatic, that is if it is used to guide experimental and theoretical inquiry within the domain of the paradigm, knowledge of possible set of models of the theory is of primary heuristic importance in conducting this inquiry. Among these set of models will be descriptions of subdomains and particulars in the domain of the paradigm. Equally heuristically important for observation and other inferences is knowledge of possible set of models in related domains that are already known or accessible to observation. Knowledge of both central and peripheral set of models can be used to guide experimental activities and strategies for observation. Also among these set of models are ideal epistemic worlds u^+ in which not only are the particulars and subdomains described, but also the paradigmatic laws are known. However, the demand to know laws cannot be deduced in PIE-QIE+ because the set u^+ of epistemic ideal worlds in which the laws are known is a subset of u of the possible worlds in which the laws hold. At best this obligation can be defended only as heuristic priority and a component of general metascientific principles based on prior scientific experience.

Knowledge of laws is primary as a presupposition of Lakatosian positive and negative heuristics, which, in turn, are based upon historical reconstructions: To deliberately and reasonably protect the 'core' of laws and other assumptions central to a scientific research program, as prescribed by Lakatos' negative heuristic, one must justifiably claim to know them. Likewise one must reasonably claim to know the core in justifiably employing it to gain theoretical and empirical knowledge of subdomains of the core and of individuals in those domains. Thus embracing both 'negative' and 'positive' heuristics with good reason presupposes that one is satisfied with her search for knowledge of the nomological content of the core. If anything in a research program is prior to further investigations within that program, it would be knowledge of the laws presupposed by the program, and in this case one might be heuristically justified in claiming, as suggested in an earlier chapter, that knowledge of such laws is demanded in the program.

A question that has not yet been addressed is how to formulate questions requesting knowledge of laws. The expression

$$!K{<}i{>}(x)(Px \twoheadrightarrow Qx)$$

is of the form $!K{<}i{>}p$ and is not suitable as an expression of a question. The reason is that by (T3+) these expressions entail presuppositions p and $(x)(Px \twoheadrightarrow Qx)$. This kind of request for knowledge of p cannot be usefully made because its permissibility can be established only if p is known to be true. Questions thus formulated are caught on the horns of Plato's *Meno* dilemma: If the search for knowledge of p requires that p already be known, then the search is either impossible, because it presupposes its own unfulfilled goal, or it is useless because the truth of its presupposition is the fulfillment of its goal. Thus the search for knowledge of p or $(x)(Px \twoheadrightarrow Qx)$ must be requested by a form that does not entail the fulfillment of the search, $!(K{<}i{>}p \vee K{<}i{>}{\sim}p)$ or $!(K{<}i{>}(x)(Px \twoheadrightarrow Qx) \vee K{<}i{>}{\sim}(x)(Px \twoheadrightarrow Qx)$, which is the standard form for asking whether p or the law $(x)(Px \twoheadrightarrow Qx)$ is true. In these formulations the presupposition is a disjunction of the form $p \vee {\sim}p$. This seemingly tautologous presupposition need not be considered always fulfilled, for a person may lack the necessary conceptual apparatus for entertaining either alternative: Aristotle cannot be expected to have asked whether Huygens' wave theory is true.

Also the above formulation of the whether question for a law is of no use in permitting the inference from a demand for knowledge of causes, $!(Ex)K{<}i{>}Px$, to a demand for knowledge of effects, $!(Ex)K{<}i{>}Qx$. If $[(x)(Px \twoheadrightarrow Qx) \vee (Ex)(Px \& {\sim}Qx)]$ and ExPx hold in all u^+, ExQx may not thus hold, for these u^+ might be a worlds in which $Px \& {\sim}Qx$ holds. Such inferences would be important in scientific research if items could be known by their effects, as many methodologists hold. If we wish to know the occurrence of something in a domain that is not directly observable, an important research strategy is to seek effects that can be attributed to this item that are directly observable. This strategy requires that we seek knowledge of a causal law producing such effects as well as of the effects themselves.

However, PIE-QIE can aid in sanctioning such a strategy if we construe the search for laws as the search for specific causal mechanisms rather than for logical relations. This construal seems to fit a plausible epistemological position regarding science: The only way that established correlations can be distinguished from accidental coincidences or from various common cause and reverse cause situations is to establish a causal mechanism that underlies and explains the correlation. A causal mechanism is a relationship among events, processes and entities meeting these general conditions, but which can differ substantively from one domain of science to another: (i) The causal conditions produce or generate, in a given environment, the effect. This is a stronger condition than found in the usual characterization of the counterfactual import of causal relations, which is limited to the claim that causal relations permit an inference from causal conditions to effects in actual and certain possible worlds. This last condition, however, is not excluded by the recommended general conditions for causal relations: (ii) Causal mechanisms delimit more specific possibilities than counterfactual conditioals and, beyond the usual stipulation for counterfactual conditions, delimit probabilities within a certain domain. Furthermore, (iii) the causes precede the effects in time, and (iv) there are discoverable properties of both cause and effect in whose terms the causal relation can be understood and distinguished from other causal mechanisms. Condition (iv) acknowledges that there are an indefinite number of different kinds of causal relation, or that different causal mechanisms can be subject to different descriptions and classifications. These descriptions can be included in the descriptive clauses of unsolved problems whose goals are the discovery of particular kinds of causal relation.

To illustrate these requirements for causal mechanisms, in the case of a process, such as a body moving under a force, condition (iv) tells us that electrical and gravitational forces are distinguished by distinct properties upon which the forces are dependent and upon which they act. Though both can be described in space as inverse square forces, the electrical force depends upon electric charge and acts thereupon and gravitation depends upon and acts upon mass. Also gravitation excludes repulsion between masses. As to condition (ii) Newton discovered and specifically described the possible trajectories under inverse square forces, viz. the various conic sections. Regarding (iii), all physical forces, including gravitation, propagate over space in finite time. In the case of causal mechanisms involving entities, e.g. species, natural selection generates species specific differences among genetically isolated populations, the latter of which follow in time the various events of the selection process. The separated populations and their respective environments have various characteristics, e.g. one population of wolves has long legs in open country and the other short legs in dense bush. These characteristics causally interact with features of the environment, and it is by virtue of this interaction that selection generates long legged and short legged varieties, subspecies and species. Different evolutionary models for different species can be expected to propose different causal mechanisms by virtue of which selection acts. Thus the mechanism for selection of melanic moths, viz. color contrast and visibility to predators, is different from that selecting long legged wolves. The possible representative species within one or more ecological regions are those allowed by natural selection over the long run, and the probable representatives in the short run are those that are best adapted to their circumstances. Excluded as possibilities would those variants whose characters persistently placed them at a disadvantage relative to their siblings, but the survival of the maladapted from one generation to the next is only improbable.

The form of a question requesting knowledge of a law may be paraphrased as a what question thus: What is the mechanism by which species are produced by environmental conditions? This kind of question can put a further constraint upon the answer in the form of a description of the mechanism, or the kind of mechanism sought: What is the *genetic* mechanism by which species are produced by environmental conditions? 'Genetic' can be expected to have different meanings in different contexts of inquiry, e.g. the Darwinian of 1837–1880, the Mendelian, the molecular. Such questions take the form !(EC)K<*i*>(C(**o**,**t**)&G(C)), where '**o**' and '**t**' name individual antecedent and consequent events or entities and 'G(C)' describes the relation sought. Requests for correlations, as in the case of Darwin's and Wallace's 'law of succession' can take a similar form except that C is described as a correlation, something meeting J.S. Mill's conditions of agreement, difference and concomitant variation. Requests for evidential relations also take a similar form: (ER)!K<*i*>[R(e,e′,h)&I(R)]. 'R' here designates an evidential relation and I(R) may include in the description the stipulation that the relation sought meets certain metascientific conditions for evidence. This form of the question indicates that the causal and evidential relata need not be the same things, the same kind of things, or in the same domain of discourse, as would be assured in the formulation !K<*i*>Px↠Qx with its single and undescribable term '↠'. The formulation of the question here advocated also leaves open the possibility of requesting interfield as well as intrafield causal relations (Darden and Maull, Maull, 1977), and requesting evidential relations between different kinds of physical entity or occurrence, such as the passage of an electron and a cloud chamber track.

Turning now to the heuristic utilization of causal and evidential relations, it might also be argued as follows for a demand that knowledge of the particulars in a paradigm's domain be produced: Lakatos' positive heuristic prescribes the extension of preferred laws, laws in the protected 'core' of the research program, to particulars in a paradigm's domain. This extension is a means to formulating the series of theories that constitute Lakatosian 'research programmes. This heuristic thus prescribes that the problem !(Ex)(Ey)(EC)K<*i*>[C(x,y)&G(C)] be instantiated for some individuals o,t in the program's field of study, say knowledge of Darwin's mockingbirds of genus *Orpheus* and species *trifasciatus*. To instantiate this problem, that is, to prescribe knowledge of a model of the evolution of *Orpheus trifasciatus*, we must insert the names '**o**' and '**t**' inside the epistemic operator 'K<*i*>'. However, this insertion requires that **o** and **t** be known. Hence to infer !(EC)K<*i*>(C(**o**,**t**)&G(C)) one must have as a premises (Ex)K<*i*>(x=**o**) and (Ex)K<*i*>(x=**t**). The method of analysis, which is a general heuristic for seeking premises needed in the desired deduction of some conclusion, in this case prescribes knowledge of **o** and **t**, i.e. !(Ex)K<*i*>(x=**o**), etc. This illustrates how the logic of PIE-QIE must be supplemented by extralogical heuristics for supporting appropriate epistemic objectives or problems.

To further investigate the significance of PIE-QIE+ for evidential relations, Clark Glymour (1980) offered a theory of evidence in which sentences e and e′ can constitute evidence bearing upon a hypothesis h if h is embedded in background beliefs B and these conditions are met: (i) The components of B are consistent with h,e and e′. (ii) There is a set of descriptions or functions $\{P_i\}$ and a set of inferences or computations R (evidential relations) from e of instantiations of these descriptions or values of these quantities such that every description or function occurring nonvacuously in h occurs among the P_i and all principles used in the R are in B. (iii) e and the values of the $\{P_i\}$ computed or inferred from e, constitutes a particular kind of evidence for h by virtue of the knowledge that h is satisfied in every structure <D,f> in which e is satisfied and in which for all d in D there is a term t described in e such that f(t)=d. In other words, there is a known principle f(t)=d by which we can infer or compute that all instances of e(t) are instances of h(d). (iv) There is a sentence e′ containing only the descriptions and variables contained in e and a set of descriptions or functions $\{S_i\}_{-}\{P_i\}$ occurring nonvacuously in h such that R includes computations of all quantities or inferences of all individual described in $\{S_i\}$ from e′, say e′(t′) and f(t′)=d′ where f(t′) is in $\{S_i\}$. e′ is also consistent with all principles in R thus used, but e′(t′) with computed values or singular terms d′ in $\{S_i\}$ are incompatible with h, or satisfy ∼h(d′).

Glymour's condition (iv) corresponds to Popper's falsifiability with the acknowledgment that knowledge of conditions of falsifiability depend upon knowledge of substantive relations R, not just upon the 'logic' of the terminology in h. This condition is a variant of the usual hypothetico-deductive requirement for hypotheses, that hypotheses be tested by observable consequences that can be drawn from them. However condition (iii), which allows deduction of instances of hypotheses from evidence by means of beliefs composing scientifically relevant background knowledge, is not usually acknowledged in the literature on empirical evidence for science. (Cf. Nickles' 'discoverability' discussed earlier). As Glymour and a few others have pointed out (Hesse, 1974, Nickles,

1985, Glymour, 1980) such deductions from evidence do occur in the history of science: Given the law of inertia (evidential relation R in background B), Newton deduced the existence of central forces acting on a planet from Kepler's second law and deduced that these forces were inverse square from Kepler's third and first laws. This kind of argument is what is sought when one seeks empirical demonstrations of scientific hypotheses (Harre, 1972). The separation of electroscope leaves indicate electric field strength in a region because the field strength (d in h) is, within specifiable limits, approximately a linear function of the angle of separation (t in e). The argument becomes highly consilient when instruments capable of imaging the object of study are employed, as can be seen in mapping an electric field with the electroscope, or even more convincingly, in representing a magnetic field with iron filings. Here d (the direction of the field) is (approximately) a function of t (the orientation of the iron filings). The coherence of the map, the approximate continuity of its contours and their obedience of Faraday's laws, provides considerable consilience among many individual indicators. Other examples of comparable imaging include Galileo's telescopic observations of Jupiter's satellites, sunspots, and the Moon's mountains, where satellite continuity, sunspot parallax and shadows cast by lunar mountains provide the consilience. Similar consilience without imaging can be had from X-ray and electronic diffraction patterns for crystals, and electrons provide consilience with images in the electron microscope.

However, though experimenters do hope to find circumstances and evidential relations in which determinate computation of hypothetical variables from observed variables is warranted, they often have to, and can make do with a lot less. The inference of h from e often involves the exclusion of a number of rivals to h, an objective that requires much creativity on the part of the experimenter because of great difficulties in making such exclusions convincing and conclusive. The experimenter's creativity is equally challenged in thinking up rival explanations to an observation to be excluded, and the search for such rivals and their refutations should be prescribed by a fundamental experimental heuristic principle. This uncertainty in support for h is often the

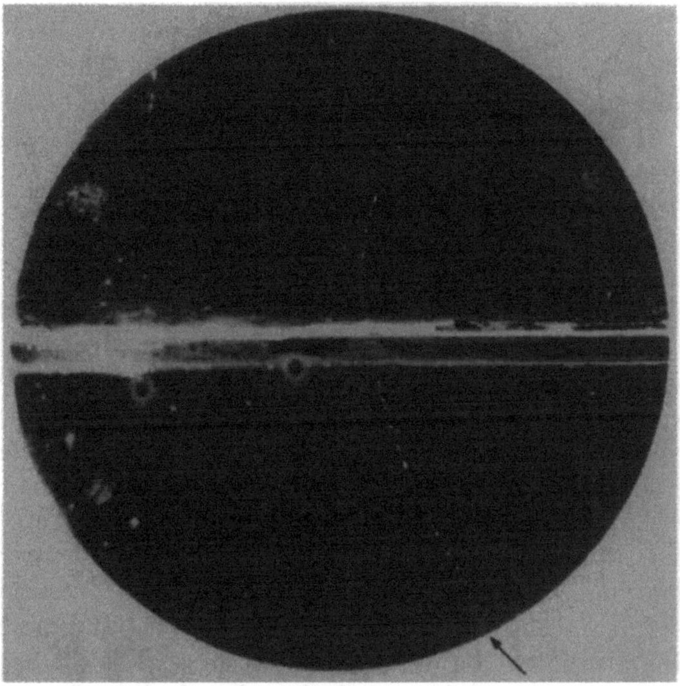

Figure 1: From Galison (1987) p. 91.
Courtesy of the Archives, California Institute of Technology.

case even in for imaging devices, such as the Wilson cloud chamber, in which one might expect the consilience just mentioned to carry decisive evidential force (Galison, 1987, p. 77). Even with cloud chamber tracks observed correlations can be accidental coincidences, the effects of common causes or reversed temporal orderings, all of which can be plausible rival beliefs about background phenomena against which the significance of an observation is assessed. Anderson's leftward curving track passing through a lead plate in an inward directed magnetic field suffers a change in curvature on either side of the plate. (See Figure 1.) The track could be coincidentally contiguous, a consequence of distinct electrons simultaneously entering and leaving the lead, two secondary particles ejected by a third (common cause), or a positron decelerating from the more curved to the less curved portion. The presence of the lead and energy conservation rules out a reversed causal process in which an electron is moving in the opposite direction, but still does not decisively support the positron hypothesis. (Galison, 1987; also Alvarez and Asaro, 1990, Courtillot, 1990 contain examples illustrating rival interpretations of correlations, e.g. between the dating of the KT boundary terminating the Cretaceous and the disappearance of dinosaur bones.) Rival interpretations of particle tracks can also include alternative identifications of the origins as well as the cause of a given trace like a particle track: Artifacts can be produced by measuring instruments, the intervening medium, or other items in the background. Particle tracks can be secondary radiation from components of the cloud chamber or the atmosphere and thus effects of cosmic gamma rays, as Robert Millikan supposed in the 1920's. Alternatively they can be primary emissions from various disintegrative processes in stars and elsewhere, contrary to Millikan's hypothesis (Galison, 1987, p. 84). One also might be mistaken in the physics of the apparatus used, and thus about Glymour's function d=f(t). In measuring energies u of cosmic rays Millikan mistakenly assumed that absorption of cosmic rays, supposed to be gamma rays emitted from cosmic production of helium, oxygen and silicon, depends only on ionization and Compton scattering. In calculating effective penetration he implicitly assumed that penetration is a function only of mass, not of density as most theories would warrant. As Robert Oppenheimer pointed out, the formula that Millikan used to determine cosmic ray energies **u**, $I(x)=I_0{}^{[-ux]}$, where I_0 is the intensity at the surface of the absorber, does not take into account other processes in absorption such as pair production. Furthermore the line between random and systematic errors may be very difficult to draw. In his oil drop experiments conducted around 1910 Millikan ignored as perturbed by uncontrollable conditions in the environment or apparatus measurements giving, for example, e=4.96 rather than the preferred e=4.77. A rival view that these errors are systematic would take the measurements as indicators of something significantly wrong with prior beliefs about the fundamental unit of electric charge. Thus it could be that there is no such indivisible unit, as held by the Viennese physicist Felix Ehrenhaft (Galison, p. 88). The energy sources used in an experiment can also generate artifacts: In efforts to measure molecular magnetic moments by rotating iron bars, magnetic fields produced by the electric motor driving the apparatus could influence measurements of the field produced by the rotating iron bar (Galison, 1987, pp. 54f). Finally Incorrect beliefs about the background environment may be the source of misinterpretations of a piece of evidence: The amount of iridium in the KT boundary of the Cretaceous indicates an asteroid impact just if the estimation of very low terrestrial iridium is correct (Alvarez *et. al.*, Courtillot, 1990).

Thus, though Glymour's conditions are epistemically optimally useful when they can be met, we cannot expect them always to be met in all circumstances encountered in scientific inquiry. Often the best we can obtain is this: given refutations of known rivals h_i to preferred hypothesis h_1, e renders h_1 most probably true. Thus placing a categorical obligation Glymour's conditions on the grounds that they provide ideal epistemic certainty would be unrealistic because scientists must and can draw conclusions that stand up in history with weaker evidential conditions. Glymour's evidential conditions would, if their realization could be construed as an epistemic ideal holding in all epistemic ideal worlds u^+, provide a deductive basis for inferring from the priority of $K\langle i\rangle h \vee K\langle i\rangle{\sim}h$ to equal priority for $K\langle i\rangle e \vee K\langle i\rangle(e' \vee \ldots)$. That is, suppose for any e,e′,h, $!(ER)K\langle i\rangle[R(e,e',h)\&I(R)]$, where I(R) describes sought after evidential relation as meeting Glymour's conditions. Then, since $K\langle i\rangle[R(e,e',h)\&I(R)]$ holds in all u^+, if $K\langle i\rangle h \vee K\langle i\rangle{\sim}h$ holds in all u^+, it can be inferred that $K\langle i\rangle e \vee K\langle i\rangle(e' \vee \ldots)$ holds in those u^+'s. Again, unless Glymour's condition is demanded for empirical evidence, this kind of deduction cannot be carried out in PIE-QIE+.

Suppose 'e_i' is taken to designate the total positive evidence for an explanatory h_k at some stage of some experimental project, which can contain many experimental operations with various apparatus substitutions and manipulations. We expect as a minimal epistemic requirement e_i, negative evidence e_j' and h_k to be so related in all u^+ that e_i, if true, will allow an h_1 to be preferred as more probably true than rival h_k's. Since we have construed evidential relations R as substantive optical, mechanical, chemical, electromagnetic, genetic, behavioral, etc. relations, probability relations among e_i,e_j',h_k should be '*natural*' probabilities (in a very broad sense including probabilities of genetic recombination, survival, behavior responses, etc.). (Giere, 1988) Natural probability is what we ascribe to a pair of presumably fair dice when we claim that the probability of throwing snake-eyes is one in thirty-six. It is also in terms of this kind of probability that we describe the trajectory of an electron passing through a crystal. However, such probabilities can only be known with mechanisms governing the behavior of specific kinds of object, e.g. dice made of homogeneous materials thrown under randomly varying initial mechanical conditions or quantum wavicles subjected to circumstances producing diffraction and interference. Also Popper is correct in claiming that all natural probabilities are dispositional: The probability of an outcome e is determined by initial and boundary conditions c within which a mechanism or an ensemble of events or processes occur, as should be clear from the two examples just cited. All physical probabilities are thus conditional and stated in the form $P(e,c)$. Finally the relations R govern certain natural possibilities: The theories T ascribing probabilities, whether to all quantum wavicles or to homogeneous cubes allowed to fall under random initial mechanical conditions, determine a set u of possible worlds. Insofar as these worlds are of interest to the inquirer, the set u will have a subset u^+ of worlds containing knowledge of various probable outcomes e under various circumstances c. Probabilistic relations R, then, will allow us, presuming the occurrence of something described by e_i, to infer that naturally it is probable that h_k is true, that is what h_k describes will occur.

If we can thus speak of probabilities in evidential relations R among e_i, e_j', h_k, then for some e_1 the probability axioms, and also Bayes' theorem in one of its several forms

$$P(h_k,e_i\&e_j) = P(h_k,e_i)\times P(e_j,h_k\&e_i)/P(e_j,e_i)$$

should hold among initial condition e_i, additional conditions e_j, and the occurrence of h_k. If e_i is considered prior and e_j posterior evidence, the denominator $P(e_j,e_i)$ is relatively small in value if the two items of evidence are, apart from the mechanisms proposed by hypothesis h_k under consideration, relatively independent of each other. This point can be seen more directly from an alternative but equivalent denominator for Bayes theorem:

$$\sum_m P(h_m, e_i) \times P(e_j, h_m \& e_i) \qquad m = 1, \ldots k, \ldots, n$$

The prior probability $P(h_m,e_i)$ and the inverse $P(e_j,h_m\&e_i)$ can be evaluated for the many h_m which include h_k but also many of its rivals. If the values of prior and inverse terms for h_k's rivals are significantly less than those for h_k, since all such values are less than or equal to unity, the denominator will be significantly smaller in Bayes' theorem, and thus the value of the posterior probability for h_k, $P(h_k,e_i\&e_j)$, is enhanced. This consequence is what we would expect from consilient evidence. However, it should be observed that consilient sources of evidence are not actually physically independent of each other. Rather, the hypothesis they support best shows how they are physically related. It shows how the occurrence of one can influence the probability of the other's occurrence. Thus, according to this epistemic interpretation of Bayes' theorem hypotheses must 'bootstrap' themselves into credibility (Glymour, 1980), that is, they must be the source of at least part of their evidential relations. Apart from the existence of surface irregularities on the moon, we would not expect Galileo's computations of elevation h of feature m on the Moon from sunlight incidence angles and observed shadows on July 14 and 15 to agree or cohere.

Also the prior probability, $P(h_k,e_i)$, directly influences the value of the posterior probability, $P(h_k,e_i\&e_j)$: e_j will further enhance the probability of h_k if the latter probability is enhanced by other empirical evidence or otherwise known circumstances e_i. Again this gain in probability is plausible on grounds of consilience. Rising inventories will enhance the probability that a recession is occurring if it is added to evidence already raising that probability, such as increased unemployment or decreased consumer confidence. Each of these pieces of evidence is a participant in recessionary cycles where unemployment or the belief that it is rising probably will decrease consumer confidence and purchasing power causing, that is, rendering more probable, rising inventories and further unemployment.

Finally the posterior probability also depends directly upon the inverse $P(e_j, h_k \& e_i)$. This feature of the inverse tells us that the most probable occurrences, as determined from mechanisms proposed in h_k and its previous supporting evidence and not from its rivals among the h_m, serves as the best evidence for h_k. According to Baysean methodologists this point applies as much to initial evidence e_i as to subsequent evidence e_j. If the mechanism proposed by h_k and rendered likely by e_i renders it physically or behaviorally more probable that e_j will occur than any other happening e'_j, then the truth or actuality of e_j will make more probable the truth or actuality of h_k. Thus a probable causal consequence of recession, such as decreasing consumer confidence, is based on the cognitive mechanism that consumers will probably become aware of possible erosion of their income when others start to lose jobs or the news media speaks of recession. As a probable consequence they will postpone optional spending on expensive items such as automobiles and homes, which, in turn will probably produce rises in inventories and probably further unemployment and bad news. If there are no other probable causes of this widespread loss of confidence it can serve as evidence that recession is probably taking place. To return to Galileo, given a background B including geometrical optics, and knowledge of the relative locations of the Sun, Moon and Earth, nothing we can think of but the existence of surface irregularities will produce shadows trigonometrically proportional to elevations h of the Moon's various features. Hence the shadows render it highly probable that the Moon's surface is irregular like the Earth's.

Glymour's requirements for evidence are special cases of a broader range of evidential relations that can be summarized in the interpretation just given of Bayes' formulae. The Galileo case meets Glymour's conditions, for we can, by means of geometrical optics, infer elevation from shadow length and incidence of sunlight. Different values of shadow length under the same incidence of sunlight will allow us to calculate different elevations that will decisively exclude certain other values for elevation. However, the recession case does not meet these conditions, for the best and most consilient evidence that a recession is taking place only enhances its probability. Also from this interpretation it emerges that evidential relations R are derived from causal mechanisms in which hypothesized events, entities or processes have characteristics whereby their occurrence *c* determines or renders probable some *e*. Thus we can assign an enhanced value to the inverse in the Bayes formula, $P(e,c)$, or if h_k describes this causal relation in some actual circumstance evidenced by e_i, and e_j describes *e*, $P(e_j,h_k\&e_i)$. Knowledge of these causal mechanisms, if not of their detailed workings or the properties by which *c* enhances *e*'s probability, knowledge at least that $h_k\&e_i$ makes e_j more probable than other rival h_m, is necessary for there to be any empirical evidence for h_k. Hence if any item in the Bayes formulae should be known as a means to knowing h_k it should be the inverse $P(e_j,h_k\&e_i)$. Thus knowledge of the inverse term should occur in any epistemic ideal world u^+ that knowledge of h_k occurs. However, if h_k proposes the causal mechanism by which e_j is made probable, Glymour's arguments show that consilience is required. We have seen that consilience can be found in enhanced probabilities for prior terms, $P(h_k,e_i)$, or in low values for the independence term in the denominator, $P(e_j,e_i)$. Hence the inverse term and either the prior or the independence term of the Bayes formula should also be known in any u^+ in which knowledge of h_k is sought.

Finally, it would appear initially that since the probability that conditions described by h_k actually occur is not determined just from e_j, but also from e_i as well, we cannot say that the obligation to know e_j transfers from the obligation to know h_k as necessary condition for the latter knowledge. If h_k can be known from e_i as well as from e_j, then e_j cannot be necessary for knowledge of h_k. Thus we cannot infer that knowledge of e_j occurs in every u^+ in which knowledge of h_k occurs, at least on the basis of our interpretation of Bayes' formula. However, since consilience is necessary to justify the bootstrapping that h_k provides for itself, we may conclude that the project of knowing h_k does require knowledge of at least two items causally relevant to the processes described by h_k, viz. knowledge of both e_i and e_j. As Glymour (1980) points out, knowledge that e'_j is a refuting condition and that either e_i or e_j is confirming evidence for h_k will satisfy the consilience requirement.

Thus Bayes' theorem allows this inference regarding the epistemic importance of items of evidence e_i and e_j or of e_i and e'_j for an hypothesis h_k: Whatever importance there is to knowing h_k can also be attributed to knowing probable consequences e_i and e_j of h_k's truth or of incompatible evidence e'_j. Whatever epistemic ideal world u^+ contains knowledge of h_k also contains knowledge of e_i and e_j or some e_i and e'_j.

It should be added that in searching for evidential relations for some h_k our search should not be biased in favor of supportive evidence. This heuristic principle can be called the Principle of Unbiased Evidential Inquiry (PUEI). Hence using PUEI we can argue that if two items of evidence are to be sought to satisfy the consilience requirement, it should be at least one affirmative e_i and at least one negative e'_j. It is thus that we can justify the fundamentally important relation between the value we attach to the solution of a problem, or to pursuing the problem, and the value of the problems of determining evidence for that solution, evidence that the solution solves the important epistemic problem. It is also from these considerations that we can infer from our valuation of the question 'Is h_k true?', $!(K{<}i{>}h_k \vee K{<}i{>}{\sim}h_k)$ to a like value for the question 'Is e_i true?', $![K{<}i{>}e_i \vee (K{<}i{>}e'_j \vee \ldots)]$, or the equivalent, $!(K{<}i{>}e_i \vee K{<}i{>}{\sim}e_i)$, where $e'_j v\ldots$ exhausts all of the alternatives to e_i. For any u^+ in which $K{<}i{>}h_k \vee K{<}i{>}{\sim}h_k$ holds, PUEI, consilience and the Bayes formula require also knowledge of corresponding inverse terms $P(e_i,h_k)$, $P({\sim}e_i,{\sim}h_k)$ as well as corresponding prior terms. Knowledge of the prior probabilities and the inverse terms are already required as part of the consilience requirement. This requirement is assured in the alternative expression of the independence condition in the denominator of the

Bayes formula:

$$\sum_m P(h_m, e_i) \times P(e_j, h_m \& e_i) \qquad m = 1, \ldots k, \ldots, n$$

where the n h_m include all of the alternatives to h_k. PUEI assures that knowledge of the corresponding prior and inverse terms for e'_j are in any u^+ in which knowledge of h_k occurs, for it requires that for every computation of the value of positive e_i or e_j in the Bayes formula, there correspond like computations of the value of negative e'_j. Thus knowledge of the following terms must also occur in these u^+: $P(h_k, e_i \& e'_j)$ and hence $P(e'_j, h_k \& e_i)$ and either $P(e'_j, e_i)$ or

$$\sum_m P(h_m, e_i) \times P(e'_j, h_m \& e_i) \qquad m = 1, \ldots k, \ldots, n$$

It is thus only in the light of requirements for evidence gleaned from studying actual cases in science that inferences from problems concerning scientific hypotheses, such as 'Is h true?', to problems concerning evidence, 'By what relation can e bear upon h?' and 'Is e true' can be warranted. Such inferences allow us to transfer the importance or the priority assigned to the demand for knowledge of h to the demand for knowledge of appropriate evidential relation R or item of evidence e.

4.4. CRITERIA FOR WEIGHTING PROBLEMS

As a means to evaluating the relative epistemic importance of questions we have considered thus far possible deductions in PIE-QIE+ and other inferences using extralogical nomological and heuristic principles, such as Lakatos' positive heuristic, PUEI and the method of analysis. These inferences allow in certain cases the transfer of epistemic value from problem to subproblem. This kind of inference is particularly important in justifying both inquiries into possible evidential relations for a given hypothesis and efforts to produce what is known to count as evidence for or against an hypothesis. However we have not considered any principles for valuing or weighting epistemic problems from the start.

Larry Laudan (1977) has offered criteria for weighting scientific problems. However the weighting he discusses is not for the purpose of the pursuit of the problem, but rather is a means of evaluating evidence for theories that solve the problems. Laudan's account of problems is limited to the 'context of justification' in the sense that he has nothing to say, at least in his (1977), about conceptualization, the construction of theories, or the search for suitable concepts for the solution of certain problems. Nor does he say anything about the search for evidential relations or problems in certifying evidence. In short, in his account ready-made theories are appraised in terms of their capacity for solving problems judged to be important. The appraisal is not as a candidate for belief, for Laudan denies that 'truth', 'probable truth' or 'proximity to truth' are important considerations in the appraisal of scientific theories. At most theories are appraised as candidates for acceptance, but not acceptance as true, that is belief (van Fraassen, 1980).

Laudan distinguishes two kinds of problem as important in the appraisal of scientific theories: (i) 'Empirical problems' are problems in which something "about the natural world", viz. facts, observed or unobserved, requires explanation (1977, p. 15). (ii) 'Conceptual problems' concern the conceptual structure of theories, and include problems raised by the lack of internal coherence within a theory, coherence relations among theories or between theories and other background beliefs, and the presence of *ad hoc* hypotheses, explanatory circularity, vagueness, and ambiguity within a theory. There is patent tension between Laudan's efforts to avoid issues bearing on the truth of scientific theories and his advocacy of explanation as one of the principal grounds for the appraisal of a scientific theory. How can a theory be accepted as having provided an explanation of some phenomenon or law unless it is accepted as true? Surely Darwin cannot explain the presence of mockingbirds on the Galapagos Islands by claiming that they migrated there unless there can be given reason to believe that the migration actually took place. Evidence for his explanatory hypotheses must also be reason to believe that they are true. Problems of explanation must thus be construed as epistemic problems in the sense that hypotheses cannot solve them unless the hypotheses are believed to be true. Such problems demand reasonable grounds for belief as a component of the problem. Thus subproblems of a problem of explanation include finding evidential relations whose satisfaction is sufficient either for truth or for inferring probable truth. In short, Laudan's position on scientific truth is highly implausible and should be dismissed from the start.

The link between Laudan's conceptual problems and truth is less clear than in the case of empirical problems. Whether 'coherence' within or among theories, non-ad hocness or clarity and precision can count toward presumed truth has long been an issue between realistic and idealistic philosophers and remains a problem in the epistemology of science. More will be said on this issue in Chapter 6 below.

Laudan has a problem in defining what counts as a 'proper' solution to a problem that threatens his whole theory of scientific appraisal. What would be an 'improper' use of T to solve P? A plausible answer would be the use of an unacceptable theory to solve P. Just solving P might be considered in itself sufficient to render T acceptable, but Laudan does not tell us enough about the demands put forth by problems to make this point plausible. Why should we value coherence, clarity or non-circularity if they are not indicators of truth? What is explanation without presumed truth? Theories might be justified as calculational instruments or crowd pleasing fictions, but Laudan has not made it clear or defended the view that the aim of science is computation or telling entertaining and somehow scientifically profound stories. Again, do scientists properly answer questions only with theories they justifiably accept? If so, how can a theory T's solution of a problem P count toward accepting T if, in order to determine that T solves P one must accept T to begin with? How could we say that the first problem P allegedly solved by T is in fact solved by T if T's evidence can only be the solution of problems?

This circularity might be avoided if the emphasis of Laudan's theory is shifted from the context of justification, i.e. circumstances in which the objective is to determine the acceptability of a given T, to a context of discovery, namely circumstances in which the objective is to pursue or solve some epistemically important problem. Suppose criteria can be provided for the importance of a problem P that are independent of its solution by a credible theory T. Suppose there are grounds for believing that the solution to P contributes importantly to our effort to realize scientific objectives whose initial conception is grounded in metascientific and cosmological beliefs M and B. Suppose also P is judged to be scientifically important on the basis of these criteria: P refers to some gap in our knowledge, e.g. a conceptual problem raised by the lack of an intelligible connection between some cause that accompanies an effect. According to Darwin's law of succession, relations of similarity among species are proportional to their geographical and temporal proximity. How are we to understand this correlation? It occurs too frequently to be coincidence, so there must be some suitable mechanism for its production. What constitutes a suitable mechanism? A hint regarding what might solve the problem can be taken from the fact that within a species we explain similarities among individuals in terms of genetic relationships. A mechanism like what produces family resemblances among related individuals might plausibly be sought to explain family resemblances among species. Does P place demands on a solution beyond mere acceptability of the solution, such as truth or actuality? In this case the answer must be yes: How can genetic relationships be effective in producing related organisms unless they actually exist? It appears that contrary to what Laudan wants to suggest, many scientific problems demand credible hypotheses for their solution, that is, hypotheses that can be accepted as true. Common descent with slight modification in successive generations will make the law of succession intelligible by fitting it into a pattern of natural relationships that would, if actual, generate families of related species. Thus the demands of P are such that only credible theories can solve it, but still, does a solution to P either initiate or add to T's credibility? Since we are not investigating the 'context of justification', as was Laudan, we do not really have to answer this question, though we will argue in Chapter 6 that solving problems of explanation alone is insufficient to support a theory. Consilient empirical evidence must be available before or after such solutions occur for explanations to enhance a theory's credibility. Our concern is directly with the importance of problems: How do scientists decide which epistemic objectives are the weightiest, those demanding immediate attention, considerable resources and permitting the shelving of other problems that compete for attention and resources?

In one further respect Laudan's list of problems is inadequate to the problems actually addressed by scientists. It appears that he is committed to the view that the aim of science is just the explanation of puzzling phenomena by internally and externally coherent, clear, non circular, etc. theories. Laudan neglects to consider that scientists seek to intervene in the world, to change their circumstances and to produce artifacts for the purpose of observation possibly among other scientific objectives. That is, they often address a certain kind of 'How-question': How do I navigate to a geographical point of interest? How do I build a thermometer that will read quickly, accurately and reliably the temperature of this process at these intervals? The importance of these questions presupposes an affirmative answer to another kind common in science: Should I travel to these oceanic islands? Should I build such a thermometer? Such actions can have epistemic objectives in that they make possible certain observations, which, in turn might be justified as objectives in view of their potential as evidence. Furthermore scientists explore, whether by empirical means or by the speculative extension of theories. Such exploration is also not always clearly subordinate to explanatory aims or objectives of finding evidence for or against some hypothesis or theory, though intelligent exploration must be selective and is thus always guided by substantive presuppositions and valuations of importance, that is by a more or less definite question. We have seen that questions require some conjecture or extension of prior belief into the unknown domain and some beliefs about relations between this domain and those that are known or are empirically accessible. Finally, it is not clear that just theories solve scientific problems, though theories probably play an essential contributing role in the solution of most scientific problems. Nor is it clear that theories always gain in truth or other value by solving these problems, particularly where it is not clear what theories are valued for besides their truth. If they are valued for truth, it is not clear that they get this value from solving problems of explanation or for having their conceptual problems solved: T may provide concepts that are clear, non circular and generate a coherent model for some fact E, but that alone adds little to T's credibility. Such an occurrence might be considered an explanation if T is already believed to be true, but then this occurrence may not add to T's credibility. On the other hand, to distinguish T from a nice piece of mathematics one should seek evidential relations R and produce conditions in which they are satisfied. In this case a problem of explanation can be solved by a theory not previously credible, but T only plays a partial role in the solution of the problem. R and observations of E are what render T credible in this case. As we have seen in the previous chapter, the best R are unbiased toward positive and negative evidence for T, and the actualization of positive evidence should provide reasons for believing that just T is true. Solving a problem of explanation for some instance does not alone provide evidence for the exclusiveness of T's application in this instance.

According to PIE-QIE+ questions can be appraised in the same modality as propositions or statements, viz. truth and falsity. Since a question logically entails a proposition, its presupposition, it can be false if its presupposition is false. Otherwise the question can be considered to be true. However, since questions also require action and the expenditure of resources, among the true questions heuristics demands that we set priorities. Pursuit of all true questions is not humanly possible, even with the aid of massive computing power and massive federal and commercial support of science. A second mode of appraisal for questions is thus needed for heuristics, so, in addition to truth and falsity we can add relative importance to the appraisal of questions.

To consider an example, the progression from the belief that speciation takes place through reproduction to the question, "By what mechanism?" might be taken to illustrate a 'Principle of Question Propagation' (PQP) such as is attributed to Kant, viz.

"The answering of our factual (scientific) questions always paves the way to further yet unanswered questions" (Rescher, 1984, pp. 28f). We have seen in this case that Darwin's answer excludes creationists' questions by changing background beliefs B in which lie presuppositions for why-questions. However, the semantics of PIE-QIE+ excludes any logical entailment between acceptance of a genetic hypothesis for speciation and the request that a particular mechanism for genetic speciation be known.

Also, Darwin's eventual response to demands for mechanisms includes theories of isolation, inheritance of acquired characteristics, sexual and natural selection. According to PQP these answers should give rise to further questions, though which specific questions remains to be determined. Among the further questions that might be asked would be requests that particular instances of these processes be conceived and shown to occur by observation or experiment. Other questions might be raised about the sufficiency of selection under geographical isolation for speciation, and questions about the manner in which variations are inherited, blended or spread through a population, or about how they can dominate tendencies to reversion. (Darwin discusses all of these questions in the *Origin*. See Kleiner, 1979, 1981, 1985b, and 1988 for further details regarding various aspects of Darwin's program of discovery.)

It would appear that certain contingencies, viz. certain facts and Darwin's knowledge of these facts, play a role in raising these questions, e.g. the discovery of sympatric sibling species, very similar but distinct species with overlapping territories, such as Darwin's two species of *Rhea* (South American Ostrich) in Patagonia. For Darwin these facts with beliefs about the geological history of the region raise questions about the adequacy of geographical isolation as a causal factor in the explanation of speciation. Plausibly this question is raised in the light of an epistemic demand that causal hypotheses, such as speciation by isolation, be extended to all phenomena within a subject domain, in this case biogeography. Also apparent blending on some occasions of crossing, the occurrence of reversion in feral domestic species, and the appearance of atomistically inherited and recessive traits on other occasions all indicate difficulties in the selectionists' assumptions that heritable variations are 'uniform' in their occurrence, that is small and frequent, and can be preserved in a population. In this case there appears to be a lack of coherence between blending inheritance and the assumption that evolution can occur naturally through the accumulation of small heritable variations. This tension raises a conceptual problem that is acknowledged by Darwin from very early in his search for a genetic theory of speciation (DeBeer, 1960, B30–33) and is exploited in full by Darwin's critics (e.g. Fleeming Jenkin's review of the *Origin* reprinted in Hull, 1973) and his successors (Provine, 1971). One path to its resolution would be more clarity and detail about the circumstances under which speciation takes place. How under known geographical circumstances can populations be separated long enough to diverge without blending to the point at which crossing is no longer possible?

In sum, Darwin's questions are not just automatically raised or inferred in the light of answers to earlier questions. Rather, they are generated from methodological desiderata, e.g. that hypotheses should apply generally within a domain. Darwin is explicitly aware of this desideratum:

> The line of argument often pursued throughout my theory is to establish a point as a probability by induction, & to apply it as hypothesis to other points, & see whether it will solve them.— (DeBeer, 1960, D117)

In other words, unfulfilled metascientific conditions, such as explanatory adequacy to a subject-matter, internal and external coherence, suitable specificity and precision, can give rise to research problems in the context of discovery, as well as serve as sources of objections to already formulated theories, background beliefs and methodologies. (See Kuhn, 1970, on problems raised by apparent falsifying instances, Lakatos, 1970, on negative and positive heuristics, Laudan, 1977, on unsolved problems, Feyerabend, 1974, and Kleiner, 1983, on the guidance provided by suspended metascientific desiderata.) This progression of questions represents the historically important process by which Darwin's theories 'opened' and sanctioned inquiries into ecology, ethology, genetics and fossil comparative anatomy, as Darwin also notes:

> ...My theory would give zest to recent and fossil Comparative Anatomy; it would lead to study of instincts, heredity and mind heredity, whole [of] metaphysics. —It would lead to closest examination of hybridity, —to what circumstances favour crossing and what prevent it; and generation, causes of change in order to know what we have come from and to what we tend, this and direct examination of direct passages of structure in species might lead to laws of change, which would then be [the] main object of study, to guide our speculations with respect to past and future. (De-Beer, 1960, B228)

Courtship rituals and breeding habits might easily be modified in isolated populations, and where they differ sufficiently they can prevent the blending of two populations. Thus ethology can be a means of achieving the clarity needed in accounts of speciation for them to be coherent with beliefs in genetics. Darwin's mechanisms of selection and isolation thus warrant transferring the epistemic importance of a theory of evolution to problems about species specific behaviors, habits, instincts and learning processes.

Given that epistemic demands for coherence and sufficient clarity and precision for explanatory applications and empirical tests are codified in metascientific beliefs, we still might ask whether the demand for explanations and details is never-ending. Is there some point at which one can reasonably conclude that she has either arrived at a fundamental mechanism or that her conceptual specifications are sufficiently clear, distinct and precise? PIE-QIE+ does not prescribe standards for explanatory adequacy, coherence, non-ad hocness, clarity and precision or empirical adequacy. It is not clear from *a priori* conceptions of these desiderata how far their demands should be carried. Some, e.g. Rescher (1984, p. 34) and Kuhn (1970, p. 42), have suggested that stopping inquiry at some preset measure of clarity, etc. is unscientific. However, these suggestions seem not to take into account that scientific decisions must fit into an economy in which limited resources, such as various talents, the use of expensive instruments and the use of energy must be allocated among competing problems of both pragmatic and epistemic importance. Knowing a physical constant to one part in a billion in some settings may be less epistemically important than knowing it to one part in ten thousand, say in a setting in which predictions from rival theories differ more than the latter interval. So also sharply defining a species concept that a decisive conclusion can be drawn about the species status of locally variant groups of Brachiopods (Darwin, *Origin*, Chapter 11) may be less important than being able to show that the various South American mockingbirds, like the two *Rhea* species, are specifically distinct. Geographical circumstances and differences between these locally representative mockingbird species were such that their species status is crucial to deciding whether or not species transmute. The occurrence of transmutation is an important question in Darwin's time because it plausibly challenged established core beliefs in the fixity of species. What made the challenge plausible was that Darwin saw a possible means of establishing transmutation, through the study of geographically representative species. Sufficiency in clarity, coherence, precision, etc. is thus to be determined in view of the demands of important problems. It is also entirely possible, and any appropriate question methodology should leave it open as a possibility, that these appraisals of sufficiency can change. They change with shifts in the appraisals of problems. It would be unscientific to permanently exclude interest in increasing quantitative accuracy or clarity of concept beyond a point now considered satisfactory. It is also heuristically unacceptable that the efforts at producing such increase should draw large resources and proceed unchecked as ends-in-themselves.

Contrary to suggestions by Laudan (1977, p. 18) and Rescher (1984, p. 48), a suggestion that is also prominent in Popper's critique of inductivism (1958, 1972), unsolved problems can be and are appraised on various grounds in science independently of the availability of theories for which their solutions might count as evidence. This appraisal is an essential component of the scientific enterprise if it is to be construed as consisting of at least potentially rationally organized and justified research agendas, as opposed to undirected random exploration. Also Rescher's argument

> The crucial fact is that progressiveness, insignificance, importance, interest and so on are all state-of-the-art relative conceptions. To apply these ideas we must already have a scientific corpus in hand to serve as a vantage point for their assessment. No commitment-neutral basis is available for deciding whether S_1 is progressive vis-a-vis S_2 or the reverse. If the test of a theory is to be its problem solving capacity —the capacity to provide viable answers to interesting questions— then this is something in which the theory itself (as an integral component of their scientific "state of the art") is going to play a pivotal role. And this is not an ultimately happy state of affairs because of the circularity it engenders. (1984, p. 48)

shows that various components of the scientific enterprise contain or presuppose a body of substantive belief about the subject matter of inquiry. It is because of the apparent circularity thus engendered for problem epistemologies that Rescher retreats into an epistemological pragmatism regarding the progress of science. According to Rescher, scientific progress should be measured in terms of accumulated powers of prediction and control rather than by an epistemic criterion, such as descriptive adequacy to a subject-matter or adequacy to a presumably constant field of problems generated or supported by purely epistemic concerns.

However, Rescher's circularity problem may be resolved by noting that the scientific 'state-of-the-art' is represented in a relatively, but not entirely stable cluster of ontological and metascientific beliefs. There may be tensions between the state-of-the-art thus understood and beliefs about what theories or ontological beliefs should be. That is, background beliefs may fall short of fulfilling metascientific ideals. They may be insufficiently specific, their application to known items may be in doubt or yet to be produced, there may be incoherences among them or they may be without attainable direct or indirect empirical support. The theories sought in most scientific investigations, particularly those falling under Kuhn's 'normal science', attempt to specify, clarify apply and ramify some of this background ontology. Epistemically successful models or their implications may be introduced into this background and may replace some of its components. Since these contributions may contradict the *status quo* in the various ontological, metascientific, or empirical components B,M,R, or E in more or less profound ways, revolutionary changes in science are marked by significant changes in one or another of these components.

The following may be considered as principles underlying the heuristic ordering of scientific problems:

1. A problem can be strongly favored on grounds of its 'archetypal' import (Laudan, 1977, pp. 34,36). This dimension of importance is based upon perceptions of its potential for undermining or decisively supporting some 'archetypal' principle underlying a scientific research tradition or research program. To put this point more generally and precisely, if a problem either threatens or promises to decisively support some element of the ontology, metascience, empirical relevance relations or conceptual systems upon which a research program is founded, it becomes particularly urgent for participants in the program or discipline. Rescher thus suggests with particularly evocative language that

> The 'importance' of a factual question Q ... turns in the final analysis on how substantial a revision of our body of scientific beliefs S is wrought by grappling with it —the extent to which answering it causes geological tremors across the cognitive landscape. (1984, p. 45)

Problems that can be ascribed to the conceptual, ontological, methodological or substantive core of a scientific research program gain importance by virtue of the generality of their ramifications.

(Cf. Laudan, 1977, 'archetype construction', p. 34, and 'archetype modification', p. 36.) Thus ascribing problems is claiming that a particular deficiency in regard to metascientific beliefs is attributable to some specific element of the ontological background, or to the conceptual system in relation to some element of the ontology, or to a theoretical model that employs established concepts either in relation to the ontology or in relation to some evidential relation that is essential to the research program.

The import of 'archetypal problems', as Rescher suggests, lies in the expectation that their solution can have an impact on a relatively large number of particular theories and experimental designs with metascientific, cosmological, conceptual and empirical presuppositions. Problems that bear upon the entire cognitive fabric of a scientific research program should be of high heuristic priority because of the relatively extensive bias that archetypal conceptions can introduce into the research enterprise. The threat of error becomes particularly severe because of its cost when the enterprise requires a relatively large investment of talent, time, equipment or energy. To put this point more optimistically, a problem pointing to a fundamental difficulty can be highlighted in importance because its expected solution might cast new insight upon a wide range of unsolved problems. Such insight can require more fruitful reformulations of the problems as well as provide novel heuristics or sources of evidence for further inquiry. Thus Darwin defends his evolutionary mechanisms in part by appeal to the 'zest' that they will give to fossil comparative anatomy, ethology, etc. This point may underlie Kant's principle of question propagation.

Archetypal problems can be made even weightier if they bear upon what are construed as more fundamental theories. Problems of atomic physics or cosmology enjoy such great epistemic importance that large financial investments seem justified in ever more powerful accelerators for testing or deciding between rival fundamental particle theories. The view that some subject matter S is fundamental, or that S is more fundamental than some other S′, is analogous to the belief that some principle or law is at the core of a research program. Significant changes in beliefs about S can have wide ranging implications regarding concepts and theories describing higher levels of organization, as in the case of the effect of molecular theories upon genetics and physiology.

A practical ramification of this criterion of importance appears in the anticipation of the great powers of intervention that can be provided by knowledge of fundamental processes in both atomic physics and genetics. In 1916 the geneticist H.J. Muller noted the parallel between transmutation of the elements and genetic mutation and speculated about the powers of human intervention in nature were we to be able to control and produce mutation in living organisms as well as in material substances at will. Thus humans possibly can gain control of the characteristics of any biological individual and the evolutionary process as well as knowledge of the fundamental processes of life. (Keller, 1990, pp. 397f). Theoretical considerations for importance are also offered by both Muller and Erwin Schrodinger, who attribute great physical as well as biological importance to the understanding of these properties of genetic material: Autoattraction in the alignment of chromosomes, autosyntheisis in the replication of genetic material, and the consequent maintenance of genetic specificity in seemingly physically hostile contexts are quite contrary to what can be expected on grounds of received physics, such as statistical mechanics which seems to entail universal tendencies to decreased order (Keller, 1990, 401f). Muller recognizes that biological methods, such as methods of selecting pure strains and crossing them practiced by Mendelian geneticists, are inadequate for studying these properties of genetic material, but foresees that X-rays will not only serve as means of intervention by inducing mutations but also as means of observing the structure of biological molecules. Hence the problems for physics and biology seemingly raised by the properties of genetic material are highlighted by a promising though yet to be implemented means of solution.

2. Early in his discussion of problem solving Laudan introduced a qualification designed to discount the epistemic importance of 'unsolved problems', by which he means problems that to date have not been solved by any theory (1977, pp. 18–22). This qualification

is designed to avoid prescribing the easy refutation of entrenched and valuable theories, a problem with Popper's methodology that also concerned Kuhn (1970) and Lakatos (1970). As is the focus of most of Laudan's (1977), this problem arises entirely in the context of justification, that is in situation in which decisions must be made whether to continue to accept certain principles and explanations. However, in the context of discovery, that is in situations in which decisions are made about implementing the heuristics of a program in order to obtain some projected epistemic state judged desirable with varying degrees of urgency, this principle would severely limit avenues of research that have frequently generated significant if not revolutionary changes in scientific belief and inquiry. Kepler's problems of explaining the periods, distances, numbers and eccentricities of the planets, Darwin's problem of explaining the law of succession, Bohr's problem of explaining the dynamic and electrodynamic stability of the Rutherford atom, and many others amply illustrate that hitherto unsolved problems can have profound epistemic importance and should be addressed by scientific inquirers, if not given primary epistemic importance. This point should hold even for problems whose solutions are initially insufficient to destroy the credibility of foundational beliefs for a research tradition.

Yet an unsolved problem P′ in a given research program may be highlighted for participants if its solution is or has been produced in a rival program. This is not to claim that solution is a necessary condition for problem weight, as Laudan has suggested, but rather only that P′'s urgency is increased if it is solved by the rival, at least for participants in the program that is failed to solve P′ who still care to defend that program on grounds of its problem solving promise or achievement. Solutions to P′ by rival programs threaten the program by suggesting that its failures are due to some error in background belief or some comparative weakness in its empirical or conceptual heuristics. Rivalry between competing programs is an effective heuristic motivating refinements in both theoretical model and empirical information. A theoretical interpretation of some phenomenon favorable to one program can motivate its rival to refine or extend knowledge of that phenomenon or its interpretive theory in its own favor. Such strategies are exemplified in rival theories of mass extinction (Alvarez, *et. al.* (1990), Courtillot (1990). The study of such rivalries should bring out important aspects of 'dialectic' methodologies and heuristics, and it seems that dialogs between defenders of rival programs in a field is the most promising application of dialectic models of scientific inquiry (Cf. Jason, 1988).

3. Kuhn (1970) and others have pointed out that some shifts in research program or tradition in science do result in the solution of problems hitherto unsolved, but in some of these cases problems solved by older programs remain unsolved by the newer ones. Such circumstances differ from those just described principally in the temporal ordering of the discovery of problems and their solutions. In the first case a solution to P' is discovered by a rival that cannot be produced by the received program, where the rival and the received programs largely overlap in their achievements or solutions to problems $\{P\}$. In the second case a novel program becomes plausible because it solves some P' unsolved by the original program but the novel program cannot as yet solve other P'' solved by the original program. Both cases thus involve comparing programs on grounds of their problem solving achievements. The latter circumstances can thus place the same urgency on the unsolved problems P'', in this case for the upstart program. The defense of the received program will highlight problems P' that it has not solved but have been solved by the new program. In short, epistemically offensive strategies as well as epistemically defensive strategies can highlight problems.

Kepler's conceptual research leading to his three 'laws' of planetary motion is motivated by an epistemic offensive. His aim is to equal, if not improve upon the computational achievement of Copernicus' and traditional programs in planetary astronomy by using a computational theory which he believed to have solved problems of making planetary motion intelligible where the tradition had disastrously failed. He searched for computational procedures that embody his dynamical theory of causation for planetary motion, a revolutionary theory that attributes two forces to the

Sun which propagate through space and drive the planets. (See Kleiner, 1983b, for details.)

4. Problems of empirical determination, viz. problems whose epistemic goals require observation or novel relations of evidential relevance, can be highlighted as especially important on grounds of the belief that their solution will contribute evidence bearing on the credibility of some theoretical model, or even some core principle in the foundations of a research tradition. We have seen in an earlier section how by means of requirements that evidence be consilient and unbiased the import of a problem can be transferred to determining evidence for its solution.

In his search for a genetic mechanism of transmutation Darwin presupposes the requisite evidential relation for an application of the *vera causa* procedure to the problem of finding a genetic mechanism for speciation. As is well known, in his search Darwin drew extensively upon observations and putative empirical laws governing optimal practices for plant and animal breeders. The bearing that these observations have upon what happens in the wild is based upon the background assumption that the same genetic principles should govern respectively all sexually or asexually reproducing organisms. Sexually reproducing organisms form a natural kind throughout which genetic laws should be generalizable, and this natural kind encompasses wild as well as domestic animals and plants. Also for Darwin this natural kind includes humans, and thus knowledge of human genetics and demography can be extended to wild animals and plants in the same way. It is thus that problems of discovering genetic laws are highlighted for Darwin in the period in which his discovery of natural selection took place, between the spring of 1837 and October 1838 when he perceived the significance of Malthus' population theory for his search. Information about breeding and other genetic experiments become particularly important for Darwin at this time, as is evidenced in the 'Notebooks on Transmutation' by many notes and queries about observations of crossing and selection by breeders. His inquiries into human demography are similarly motivated.

Empirical problems of this kind may also be highlighted in a more ‘normal’ context of scientific inquiry in which a conceptual system, background ontology, metascientific goals and empirical methods are given and complexes of testable models are sought for the purpose of explaining known facts or laws or for discovering the unknown residue of some factual domain. We have seen how the heuristic method of analysis can transfer the importance of achieving explanations in a research program to the importance of knowing individuals or subdomains in the terms of that program. Theoretical models are the means by which these items are thus known. On the *Beagle* voyage Darwin applies the geographical-ecological species concepts that he learned from Lyell’s *Principles of Geology* in identifying and cataloging animal and plant species encountered on the various oceanic islands and continents he visited. The questions he pursued about, e.g., the geographical range of the various species he described are given structure and importance by Lyell’s conceptual system, and we have already seen that these questions did not demand sufficient specificity to serve Darwin’s later purposes of demonstrating geographic speciation. Still the application of Lyellian concepts to these species is important as a means to producing Lyellian biogeographical models for explaining their locations and their ranges.

5. The highlighting of an empirical problem as a ‘crucial experiment’ is much better known to philosophers than the other strategies for problem preference just presented. A problem to be solved to be solved by experimental manipulation, such as the measurement of the velocity of light in refracting media, may be expected to provide evidence bearing upon two or more rival claims about a subject, such as Newton’s and Huygens’ claims about the velocity of light in refracting media. These rival claims are important to these two theories of light because each plays an essential role in an paradigmatic application of the physics of light, the deduction of a theory of refraction. These problems are highlighted in proportion to estimates regarding whether their solutions will contribute to the resolution of current controversies. One could offer a strong defense for experiments designed to measure the velocity of light in various media on the grounds that it could decide between

Newton's corpuscular theory and Huygens', Young's and Fresnel's wave theory of light. The problem is perceived as promising to aid in an epistemically important decision, in this case a decision regarding a fundamental process in the Universe, the propagation of light. The planning and relatively large effort directed to an experiment accurate to $(v/c)^2$ at the end of the nineteenth century similarly can be defended as a means of deciding between various theories of the ether. Problems of this kind can be crucial to a decision between less fundamental theories, in which case their epistemic weight may be less, in accord with a weighting principle mentioned earlier.

Crucial problems need not demand observational resolutions. An application of Newton's third law to a two body system consisting of the Sun and a planet, such as the Earth or Jupiter, can be considered crucial in deciding the issue between heliocentricist and geocentrist astronomers. In this case the computations show that the Sun's motion is too small to be observed, and hence the strength of the argument in deciding the issue is based on Newton's other achievements, such as the deduction of Kepler's laws, not on an observation. At least in retrospect, this problem has profound importance in providing the strongest evidence for the Earth's motion available in the 17th century.

6. There is a third sense (the first being the relation of archetypal problems to subordinate empirical and conceptual problems and the second the relation between a problem and its subproblem) in which problem generality can bear upon its epistemic importance. For example, one might ask for a theoretical determination of some formula that will, when suitably specified, describe the motions of all harmonic oscillators. Such a demand represents a more general objective than, say, determining the motion of a taught wire under certain initial conditions. Its enhanced value is based upon an estimate of the epistemic payoff of a solution, particularly the expectation that the solution will produce or aid in the solution of other problems. Sylvain Bromberger (1985) called the means for valuing such generality 'value multipliers'. The solution to this more general problem is thus greater epistemic progress than the

solution of one or a few of a finite number of particular problems concerning harmonic oscillators.

7. Laudan has claimed that a problem can become more important the longer it remains unsolved. This point needs qualification, for epistemically and pragmatically trivial problems seemingly should remain so unless circumstances change to enhance their value. On the other hand, it does seem plausible that an unsolved important problem, such as problems raised by the precession of Mercury's orbit for newtonian celestial mechanics, should become even more troublesome if it remains unsolved for a long period of time. Still this increase in importance is plausibly a change in circumstances and an accompanying change in the expectations that scientists have with these circumstances. Scientists and philosophers expect that over a period of time the experimental, conceptual and heuristic capabilities of a research tradition should improve, or if research traditions are replaced the capabilities of the members of a discipline should improve. Intransigent problems suggest that some of these expectations may be disappointed, or that there might be some heuristic shortcoming in the research tradition or a recently initiated research program that hinders its progress. Both defensive and offensive strategies would highlight the intransigent problems under these circumstances.

4.5. MEANS AND ENDS: HEURISTICALLY UNSOLVABLE PROBLEMS

8. Problems may be highlighted in their importance when a means to their solution is at hand, or there occur indications that the development of these means is practicable. There should be little motivation to allocate precious resources to a problem that for one reason or another appears to be intractable. Giere (1988) noted that individual scientists often gain reputations for credibility by developing specialized talents and resources. When they encounter a problem that appears particularly likely to be solved by means available within these resources, whether or not the problem lies within their field, the problem can become highlighted as an object of attention.

Thus certain epistemic objectives may be believed unobtainable because it is believed that there is no means of achieving them. They might be called utopian (Laudan, 1984) and thus be considered unworthy of pursuit. Still we may distinguish those goals whose fulfillment is incompatible with background assumptions about a subject matter from those not thus incompatible. The former correspond to invalid or false questions, which need not be further considered, at least without specific reason for questioning background assumptions, but the latter can be counted among allowable desiderata for problems. Problems whose presuppositions are believed true may yet be considered insoluble because it is thought that procedures necessary for their solution, viz. subproblems and heuristics, whether experimental, observational or conceptual, are impossible either on substantive grounds or on grounds of supposed technological insufficiencies. The heuristic for solution may be excluded by some received law or application thereof, as optically observing viruses is excluded by laws and circumstances governing the resolving power of optical instruments and beliefs about the size of viruses. This exclusion does not mean that a solution to the problem of imaging viruses is physically impossible, a conclusion that would falsify the problem of producing such images. Quantum theory falsifies problems of determining simultaneous positions and momenta, etc., within the limits imposed by the Heisenberg principle; thermodynamic falsifies the problem of how to build a perpetual motion machine. This kind of heuristic shortcoming means only that a certain means to solution is physically impossible. Alternative means are not ruled out by the physical impossibility of one such means, though such alternatives may be unknown at a given time. A discovery, such as the discovery of X-rays, may provide the promise of novel means empirical access to a domain, such as the atomic structure of crystals. On the other hand, the inability to experimentally study some biochemical process in the living cell without killing the cell, and thus disrupting the process, is, unlike the limitations imposed by the Heisenberg indeterminacy principle, an example of a technical limitation on the experimental activities of biologists at one point in history. Overcoming such limitations may, but need not, require the discovery of new laws or even new observables, such as radioactive isotopes of substances known to take part in biological processes, by which such processes may be prepared for observation or by which the processes may be observed without fatal intrusion.

In cases of heuristic insufficiencies a choice can be made to leave the problem unsolved or to invest resources in an effort to overcome them. Dewey's suggestion that "... the *idea* of the object of desire and interest, the *end in view* as distinct from the end or outcome actually effected, is warranted in the precise degree in which it formed in terms of these operative conditions" (1939, p. 29), like the positivistic view that an acceptable empirical concept requires a method of testing its application, is too strong to allow the important scientific enterprise of seeking such conditions and methods for a given proposition or projected item of knowledge. Dewey's principle would rule out the search for methods of early detection of various cancers, an enterprise that is sustained by the anticipated value of its success and by a track record of the success of similar enterprises in the past. As to technical insufficiencies, investments can be placed into a search for means, e.g., to extract enzymes, to study processes in cell-free environments, or to label substances with radioactive markers for tracing their progress through living biochemical pathways. Two considerations bear prominently upon this decision to invest: (i) Knowledge of laws linking the subject of inquiry to observable domains, viz. evidential relations, can provide substantive and heuristic grounds for a new experimental technology. The technical problem of using these laws for experimental purposes would include finding or preparing circumstances in which the laws apply and in which various conditions of the object of investigation has a uniquely predictable causal influence on the measuring system, and then finding a way of observing this state of the measuring system. (ii) When no such laws are known the epistemic importance of the problem can still justify efforts to discover them for evidential relations. One heuristic for searching for evidential relations is by applications of known conceptual systems for the purpose of deriving them from models of combined observed and measuring systems. As in the case of X-rays, the potential for application to experimental and observational problems can highlight the importance of an effort to elaborate models linking observable to unobservable domains. Thus Sir Lawrence and Sir William Bragg between 1915 and 1930 applied wave optics in the search for three dimensional models of crystals as diffractors of X-rays via methods of Fourrier analysis. History has also shown that empirical evidential relations can also be sought by inductive means, e.g. by means of an application of Newton's *vera causa* procedure that allows observed causes to be supposed on grounds of similar observed effects in parts of a domain where the causes cannot be observed or detected by experiment.

The priority of these efforts thus depends upon the importance of the problem, estimates of technical and nomological possibilities of means to its solution, and estimated costs of the procedures in a context of other more or less pressing epistemic problems. The physical impossibility of directly or indirectly observing some happening does not necessarily exclude its occurrence or the importance of knowing its occurrence and can only be decided in relation to known evidential relations. This kind of impossibility thus does not imply that the happening is not problematic or that means to its observation should not be sought. Especially promising means to its solution might be evident from other empirical relations and interdomain relations that are implicit in the background ontology. Though black holes cannot be sources of light and thus cannot be directly observed, they do interact gravitationally with surrounding matter, that is, they can capture visible stars in orbits and ingest matter from them. If effects peculiar to the latter occurrences can be identified, the location of a black hole can be detected. Promising empirical relations can also be suggested by trends in producing theoretical models or the successful modeling of analogous domains. DeBroglies' wave mechanics for the electron suggested that the wavelength of commonly available electrons is comparable to that of X-Rays. Thus electrons could be used for crystallography and models of their interaction with crystal planes analogous to the Bragg's X-ray diffraction models could be expected. Unlike X-rays, magnetic fields might also be useful in focusing electrons analogously to the focusing of visible and ultraviolet light for imaging where in the case of the electrons very short wavelengths might permit the imaging of objects beneath the threshold of optical visibility.

4.6. SUMMARY AND CONCLUSIONS

1. In scientific research programs valid epistemic problems or demands for epistemic objectives can be further ordered as to importance or priority. Erotetic logic provides logical relations for the falsification of problems, but it does not provide reasons for preferring those problems that are compatible with received belief. Also paradigms and research traditions may prescribe priorities in problems, but such prescription is insufficient as a rationale for problem preference because revolutionary preferences can arise for good reason and problem choice should not be a matter of current convention.

2. One ground for priorities among scientific problems is heuristic: For example, Lakatosian heuristics attaches priority to knowledge of laws. Lakatosian positive heuristic presupposes knowledge of laws in a domain and attaches priority to the application of laws to individuals in production of Lakatosian 'theories' (models), and a means to having these models is knowledge of individuals. The method of analysis attaches equal priority to knowledge of individuals. This heuristic is a means achieving a progressive research program, where a such a program is epistemically productive, that is, it is heuristically valuable in the search for knowledge of some subject. But the program is heuristically valuable because of the value of the products, which are theoretical models of individuals and subdomains in that subject.

3. Bayes' formula allows us to infer that probable causal consequences can be used inferentially in supporting hypotheses. This inference is under the proviso that consilient evidence is found, where such evidence is demanded by the principle of unbiased evidential inquiry, PUEI. This principle requires that in searching for

evidential relations we give unbiased consideration to both possible supporting and possible refuting evidence. Thus we must know the prior and inverse terms in the Bayes' formula for both supporting evidence e_i for some h_k and evidence e'_j for rivals h_m, which would be refuting evidence for h_k. Knowledge of these terms ensure that the evidence under consideration is consilient. PIE-QIE+ is of no help in establishing problems of determining evidence as subproblems of epistemic problems, which is a fundamental relation in the methodology of scientific inquiry. Substantive causal relations that attribute at least probable outcomes e to initial conditions c and PUEI are, however, sufficient to warrant such deductions.

4. Questions were raised about the legitimacy of Laudan's program of appraising theories as problem solvers, particularly the circularity implicit in the view that what provides an acceptable solution to a problem is thereby acceptable. Contrary to Laudan's anti-realistic stance, problems of explanation in science demand evidence that their solutions describe actual occurrences or processes. However, problems can, without this circularity, be weighted not as evidence for theories but as posing valid objectives, that is objectives consistent with presuppositions or background beliefs, that are more or less important in the overall scientific enterprise. Problems that bear upon the core of received theories, research traditions or research programs, whether conceptual or empirical, are more important than peripheral problems because of the wide ranging implications of bias in the core of a theory. Analogously, problems bearing upon principles governing fundamental physical or biological processes are more important both for epistemic and practical purposes than those bearing upon less general or less essential processes. Problems that are perceived to be involved in crucial experiments or other crucial arguments that can decide an issue are highlighted in proportion to the importance of the issue, that is, according to whether the issue be about some fundamental process or at the core of some important theories. In defensive strategies problems solved by rival traditions can be highlighted within one's favored tradition. In offensive strategies problems in a novel program that have been solved by

the tradition in a discipline are highlighted. Using requirements that evidence be consilient and unbiased in the context of Baysean theories of evidence, problems concerning evidence can be highlighted by showing that this evidence is essential to the solution of a problem already taken to be important. Similarly, in the process of constructing models, the method of analysis transfers the importance of applying the concepts of a research tradition to empirical and conceptual problems entailed in the effort to know individuals or classes of systems in the terms favored by the tradition. Finally, because scientists expect heuristic capabilities of research traditions or disciplines to grow with time, persistently unsolved problems can be highlighted either in offensive or defensive research strategies.

5. The possession of technical or conceptual means to solving a problem often are grounds for selecting the problem, and many problems are left aside when it seems that no such means are accessible. However, technical difficulties and physical impossibilities need not block inquiry into means of actualizing empirical access to the subject of some important question. The promise of means to solving important problems can further highlight the heuristic importance of pursuing an already epistemically important problem, and this highlighting in turn, according to principles elaborated above, can be transferred to the importance of developing the technical means to solving the problem or to seeking whatever nomological or theoretical principle underlies the technique sought. Furthermore, the discovery of such means can heuristically highlight an already epistemically important problem that may have been put on the back burner for lack of any empirical or other means of investigation. Thus heightened interest in the physics and chemistry of life processes in the first half of this century can be attributed to promising methods of detecting molecular structural relations and methods of producing and observing genetic differences in small fast breeding organisms such as fruit flies, bacteria and yeasts.

CHAPTER FIVE

THE PURSUIT OF QUESTIONS

5.1. INTRODUCTION

This Chapter offers an account of the pursuit of problems. Since it has been assumed that scientific discoveries are the novel solution of certain important scientific problems, and scientific research is the pursuit of these problems, we are attempting an account of scientific research. We shall also be particularly interested in instruments or strategies in scientific research that guide or facilitate the solution of scientific problems.

A number of writers, viz. Newell, Simon (1972, 1977), Langley *et. al.* (1987), Holland *et. al.* (1986) and Thagard (1988) have attempted computer simulations scientific discoveries. These simulations include various attempted models of scientific research. We shall begin with a discussion of these in our effort to philosophically illuminate the nature of scientific research.

5.2. PROBLEM SPACES

The above-mentioned writers, let us refer to them as 'AI theorists', share a conception of problems that is as explicitly teleological as that we found in Hintikka-Aqvist erotetic logic, though it lacks the explicit epistemic and deontic components found there. The AI theorists' conception of scientific problems is embedded in a theory of inquiry, or more specifically, a theory of the pursuit, which is entirely lacking in the account of problems in erotetic logic. This theory was initially conceived as a means to developing 'expert' programs for problem-solving by machines, but Simon has also suggested (1977, Langley, *et. al*; 1987) means by which it may be generalized to cover much if not all scientific research.

According to Simon, a *problem space* is a symbolic representation in the brain, a kind of concept. Concepts thus characterized are symbolic representations adapted to thinking, which is information processing in the brain. By contrast an expression, such as an utterance or a written sentence, is a symbolic representation adapted to communication. Problem spaces include initial, immediate and final 'problem situations', that is cognitive states, and particularly a specific cognitive objective or 'goal state', which would, but need not, be the final state of a successful search. They also include a range of concepts employed in the 'solution process', e.g. beliefs, like background beliefs B, which constrain or guide the search for a solution. In other words, the problem space includes information presupposed by and relevant to the problem, its subject matter and its goal.

A well-defined problem space or search space, particularly one suited to the capabilities of extant machines, also requires a specific set of rules or *operators* in the problem space that apply to and thereby transform any cognitive state in the space into subsequent cognitive states. It is among the states thus generated that the problem's solution is sought and should be found (1987, p. 8). The iterated application of these operators extends the state to be explored under the constraints of the problem's goal state and relevant background information.

Simon is committed to the 'rules and representations' paradigm for human thought processes, which is not the only one being pursued in contemporary cognitive research. (See Bectel (1986, 1988) for a brief description of an alternative, parallel distributive processing or PDP and associated connectionist views of mind. The authoritative technical exposition of this approach is Rummelhart and McClelland (1986).) PDP is designed to closely model actual neural networks and thus makes use of a system or a network of simple units, that is, units that are relatively simple in structure and in their possible outputs. Their actions can be stimulated or inhibited by actions of the connections, whose activity is proportional to the actions of the connected elements. After an initial input the units interact and eventually reach a stable state of minimal energy, where the input is (or poses) the problem and the stable state the system's solution. The solution is not produced by rules or fixed operators that generate a search space by discrete sequential steps from the initial state of the system, or by solving a sequence of component subproblems by applying rules in sequence, as with the rules and representation architecture. Rather, parallel distributive processing takes the form of an overall pattern of activity constrained by a pattern of simultaneous interactions of inhibition and sensitization of many components. The relations of mutual inhibition and sensitization among the units can vary continuously and are influenced by the past history of the system's operation. The multiple inhibiting and sensitizing relations produce what is called parallel as opposed to sequential processing where in the former processes any number of operations, e.g. generating paths of exploration in a problem space, can be carried out at a time. Such processing increases indefinitely the speed with which a problem space can be generated and searched because an indefinitely large number of operators can be used at once.

Simon's rules and representations are suggestively analogous to the various conceptual systems that punctuate, divide, and sometimes coexist in the histories of various scientific disciplines. Conceptual systems are plausibly thought of as alternative modes of representation with inference rules interconnecting them. Scientific concepts C can be construed as nodes of intersection in a network of laws L, which in turn can function as rules, e.g. rules of inference and evidential relations R (Quine, 1953, Hesse, 1974, Holland, *et. al.* 1986, Thagard, 1988). History has shown us that these elements can change in a discontinuous fashion, which is not compatible with the continuity of reenforcement and inhibition postulated in PDP's network models. Thus Aristotle's laws linking applied forces with velocities can be replaced with Newton's laws that link these forces with accelerations, with consequent discontinuous implications regarding concepts of mass and weight and regarding evidential relations and the evidential bearing of behaviors of falling bodies on the possible rotation of the Earth. Also construing problems as simply activators of a neural system in PDP does not give a satisfactory account of the cognitive content of problems, which are more satisfactorily described and individuated in terms of specific initial and goal cognitive states. The goal state must be represented as a cognitive state not yet achieved by the system, a state corresponding to an ideal world in the semantics of erotetic logic, if this state is to be distinct from some actual initial state, a state in the actual world, and yet still considered or recognized as a desideratum. It is plausible that the cognitive processing evident in scientific research takes place at a different level, a level of cognition, cognitive states, problem spaces and the like, from the neural processing modeled by the PDP programs (Bechtel, 1986). However, we might also hope that an understanding of the latter level can illuminate possibilities in the former, e.g. regarding the efficiency of parallel as opposed to serial processing and regarding the manner in which concepts might be activated in the search for the solution of a problem (Thagard, 1988; see below).

There are several positions that might be taken in regard to the relation between cognitive psychology, machine inquiry and scientific inquiry. One is that the developing concepts in machine inquiry, i.e. developments in artificial intelligence (AI), will be entirely sufficient in providing models for human problem solving, which in turn is entirely sufficient to provide models for all of scientific inquiry. This program might be called the *reductionist approach* (RA). A somewhat more modest position, call it the *interfield approach* (IA) (Cf. Wimsatt, 1974, Maull and Darden, 1977, Bechtel, 1988), relates these disciplines with something less than the hope for or promise of sufficiency: Machine inquiry will aid in the solution of some problems arising in cognitive science, cognitive science, in turn, will aid to some extent in metascientific or meta-metasicientific as well as scientific inquiry, as will also developments in AI. These contributions can be expected to go in other directions as well, e.g. science and metascience, as well as reflection on variously sophisticated commonsense, have all contributed heuristics to AI research. Such aid contributes to the solution of problems in the aided field for which standing procedures in that field are insufficient, as has been illustrated in molecular genetics (Wimsatt, 1976, Maull and Darden, 1977, Maull, 1977, Darden, 1980, McCauley, 1986). Particularly interesting are borrowings of positive guidelines for carrying out research, viz. heuristics, from one field to another, in addition to borrowings of concepts.

As to a more specific defense of thesis IA, a requirement for the operators generating search spaces for machines is that they be finite iterations of algorithmic rules in which at the completion of each iteration determinate instructions are provided for the choice of an algorithm for the next step in the iteration. Rules thus linked together form 'recursive algorithms' and can be employed as 'effective procedures' for getting from initiating conditions to final outcomes in a chain of rules thus linked. The instructions in the initiating conditions of algorithmic rules could also request the input of further information sufficient to determine the next operation without themselves being thus determinate, or they could be rendered determinate just by the input of information at that stage. This constraint on the programs for machine inquiry is a general condition that machine usable heuristics and other mechanizable rules must meet. Heuristics may be construed as further procedures for restricting the search space generated by problem space operators, including the choice of 'smart' or particularly cost-effective operators, so that inquiry is aided or success can be reasonably expected. Constraints on these operators represent general architectural characteristics of hardware presently in use. As yet there is no reason for believing that human intelligence is limited to the implementation of algorithmic state space generators suited to computation by present day hardware. Furthermore technologies are as subject to revolution as is the conceptual and epistemic background of any scientific discipline. Thus the heuristic constraints imposed by present day AI programs are subject to unanticipated change. Hence it would be unreasonable to impose constraints under which contemporary machines operate upon our understanding of human intelligence in general and upon scientific research in particular. Problem-solving strategies should be investigated from several sources, including those suitable to extant machines, those that can be identified in controlled psychological experiments with human subjects, and those that can be found in the history of human inquiry, of which science is perhaps the most important component because of its relatively easily identifiable success.

Conversely, mechanized inquiry under this design paradigm may enjoy heuristic advantages beyond human capabilities, particularly when it comes to the rapid applications of iterated rules or extensive chains of rules. Mechanized inquiry of the current kind is thus always available as an aid to human inquiry. Humans can divide their problems into subproblems in which some of the tasks are particularly suitable to available procedures for mechanization. But the localized utility of mechanized procedures is not grounds for the hope that machine processes are sufficient to cover all aspects of human inquiry.

In addition to definitely specified goal states, background constraints and operators, a third requirement for Simon's WSP's is that there be an effective procedure, i.e. finitely recursive algorithms, for deciding that the goal state of a problem has been achieved by one of the states generated in the problem space. All WSP's should include the specification of a state in the problem space where the search be regarded as complete and thus allowed to stop. This desideratum is analogous to the preanalytic belief that to know a problem is to know something about what would solve the problem. Mechanizable problem-solving procedures must also include an effective and practicable procedure for identifying a state in the inquiry as a solution to the problem investigated. By the conception of problems in erotetic logic a solution to a problem is a singular term that satisfies a description, the description in the problem's desideratum and presupposition. Since the presupposition must be true before the problem can be posed and one requirement for answers is that the individual satisfying the presupposition be known, deciding whether A is an answer to question Q requires that A be shown to be true. Generally there is no recursive procedure for generating acceptable evidence for scientific propositions, even when they are answers to questions with explicit epistemic requirements. Such productions often require much beyond routinizable manipulations and adjustments of apparatus when epistemic requirements for an answer are known, and in the case of corrective answers inventiveness that is hardly at present recursively routinizable is required in the search for novel kinds of evidence. Hence this requirement that solutions be routinely identifiable is too strong for most of science as we know it today. However, the same points just made about generative operators apply here as well: Presently mechanizable decision procedures represent a state of present technology and should not be regarded as constraints that cannot be overcome. Particularly they should not delimit our views about what decision procedures are available to human investigators.

The absence of any mechanizable decision procedure is particularly evident for epistemic problems, i.e. problems whose solutions can be regarded as justifiably contributing to the fund of received knowledge. There is as yet no *generally* mechanizable way of searching for or providing evidence for scientific propositions at any level, whether conceptual, theoretical or empirical, even if the generation of propositions could be mechanized. This is not to say that in certain cases evidence cannot be generated automatically, for there are computer programs that can determine the structure of organic compounds (Buchanan, 1982) and perform medical diagnoses (Schaffner, 1985, 'Introduction'). This point is particularly evident when we remember that an important component of science is the open-ended class of substantive evidential relations R, which include guidelines for inferring actual from apparent motion, for historically interpreting fossil bearing geological formations, for inferring crystal structures from diffraction patterns, and much else. Mechanizable experimental procedures such as DENDRAL for determining molecular structures from mass spectroscopic data presuppose rather than generate such relations. Information that molecules or atoms of a given mass are of a certain kind which also have characteristic bonds and bonding configurations is built into and not generated by this program. This is not to say that some component of the search for such relations cannot be mechanized in other programs for local circumstances or even generalized with some essential restrictions.

The existence of an effective epistemic decision procedure for first order logic is well known. Here the postulates, i.e. the background B, and the rules of inference, i.e. the operators in the problem space, should be well defined in any system under investigation. Once a proof is formulated, checking its validity is a matter of algorithmic routine. This decision procedure has epistemic authority in the sense that it demonstrates the legitimacy of transferring credibility from postulates to problem solutions.

However, the view that human knowledge, or any scientific component thereof, can be represented as an axiomatic system is not generally applicable to science, as historicist philosophers have extensively demonstrated. Background beliefs B may be questioned, rejected, modified and replaced in many circumstances in scientific practice and need not be mutually consistent to play an important part in scientific inquiry. Analogs to generative operators may be found in evidential relations R from experimentally accessible domains E, but these are substantive relations based upon laws that may be questioned and revised in the process of scientific inquiry. 'Operators' for generating further steps in paths of scientific inquiry may also be found in metascientific beliefs M, e.g. Newton's prescriptions to search, e.g. in the atomic domain, for *vera causae* or causal relations for which there are already empirical demonstrations. But these causal laws too may be questioned and even replaced in plausible scientific inquiries, particularly when the inquiries extend to new domains (see Chapter 6 below). There is no fixed set of concepts C adequate to all legitimate science, as was once sought by reductive empiricists, and thus scientific inquiry in some circumstances can include the search for suitable C.

Simon's conception of 'ill defined problems' (ISP's), as opposed to 'well defined problems' (WSP's) (1977), can be regarded as foundational for the implementation of the weaker IA program for metascience. ISP's are those problems whose spaces are not clearly or decisively specified and for which there are no plausible search or decision algorithms. He says (1977) that the boundary between these problem types is vague because of the vagueness of 'practicable amounts of computation and search', a point that may reflect the lack of finality in our technical computational capabilities at any point in time. Such vagueness should not be excised by arbitrary decision. Langley, Simon *et. al.* (1987) suggest plausibly that something analogous to WSP's characterize Kuhnian 'normal science' and something more analogous to ISP's are typical in more 'extraordinary' research, where this distinction can be construed as one of degree rather than precise dichotomization. Many historicist philosophers would now agree with this last point. Also one need not subscribe to the Kuhnian historiograpical version of 'punctuated equilibrium' in which history consists of periods of conceptually stable normal science broken by episodes of revolution in order to make use of Simon's distinction. Rather, the occurrence in the history of science of problem types described in Chapter 2 suggest that revolutionary science can be characterized by the prevalence of *relatively* WSP's, e.g. problems of conceptualization, though most of these in science will not possess recursively decidable solutions. Yet logical indeterminacy of theory choice and the indecisiveness of experiments implies that normal scientific research, research in which conceptual or ontological issues are not at the center of attention, also can face ISP's. In actual science attention to ISP's may overlap the time in which WSP's are widely pursued.

The upshot of these reflections is that several positions can be taken in regard to the contrast between WSP's and ISP's in science. (i) It might be held that only WSP's are the subject matter of heuristic search and all other problems, ISP's fall outside the domain of an appropriate 'logic of discovery'. (ii) Alternatively, one might seek conditions under which ISP's can be reformulated as WSP's and thus an aspect of scientific or metascientific research might be an effort to expand the domain of a logic of discovery. Still the unreconstructed residue of ISP's cannot be subjects of systematic scientific inquiry. (iii) There may be heuristics that fall outside of the constraints of present day hardware design but nevertheless are and can be shown to be effective by some metaheuristic argument. In this case systemic scientific inquiry need not be restricted to the formulation and pursuit of WSP's, though plausibly that enterprise should be pursued as an aid to scientific inquiry particularly because of the speed with which machines can implement algorithms. This consideration, and the consideration that science has a rather large history of unmechanized inquiry should strongly support this third position, which also nicely coheres with the features of IA discussed above.

Further support of this position comes from apparent success in converting certain scientific ISP's into WSP's and reconstructing in mechanizable form certain components of the history of science. Langley, Simon *et. al.* thus claim that some ISP's can be made accessible to currently available problem solving programs whose operations are within the capabilities of current machines. Some of the successful applications of BACON in their (1987) demonstrate this point. Given the vagueness of the boundaries between ISP's and WSP's there is good reason for hoping that efforts at more precise formulation of scientific problems, at designing algorithms, and perhaps even investigating alternatives to present machine architecture will produce more wide ranging if not more powerful mechanizations of scientific research. Accordingly Langley's and Simon's program includes the following agenda item: Problem formulations should be sought that would meet stipulated conditions for WSP's appropriate to some stage in the development of a technical capability. At present this item is specified for machines that can sequentially process algorithmic programs. However, it also should be kept in mind that the successful solution of a problem can depend upon the choice of a suitable heuristic as much as upon a suitable formulation of a problem, where a heuristic is some procedure that will constrain exponentially increasing search paths to something practicable by the hardware at hand. We will see below that the PROCESS OF INDUCTION (PI) program is adapted to parallel as well as sequential processing, which has the effect of enhancing the range of problems whose solutions are practicable. But it does not appear that the form in which problems are stated need be specially adapted to parallel processing. Parallel systems at present use the same algorithms that sequential systems use, and we will see that algorithms can be construed as conditional rules in which the antecedent is an *activating condition* and the consequent an *action*. A rule can be activated by the content of a problem's *initial state*, and several rules with similar activating conditions can be activated in parallel by that initial state. Hence problems for parallel and sequential processing need not differ in their initial states. Since actions can either serve to activate linked algorithms whose activating conditions match the actions of prior rules or they can satisfy *goal conditions* of problems, and parallel systems differ from sequential only in activating several such linkages at once, the action conditions of algorithms and the goal states of problems need not differ for parallel and sequential systems. Hence the use of a parallel system is like the use of a heuristic, it has the effect of enhancing the power of a system, particularly the likelihood that it will reach a solution, and the speed of a system in solving a given problem without necessarily requiring that the problem be changed in form. This point does not exclude, however, the use of heuristics of problem revision as tools of inquiry. Nor does it imply that all tractable scientific problems should be put in the form required by sequential and parallel processors, for scientific research may be composed of activities other than the activation of algorithms that work in these machines.

In the case of scientific problems formulations 'researchable' or the more broadly practicable scientific problems as well as formulations of WSPs should be sought by specification of a suitable background ontology B. Such specifications are directed to the definition of a subject matter and relevance relations among the items of that subject and between them and other familiar, observable or manageable items of the universe. Thus the inquirer can appeal to background scientific ontologies to know what are material bodies whose mutual accelerations are to be studied, what are optical media, or what are batteries and wires. Secondly many of the historically important problems include some epistemic constraints, viz. an M specifying metascientific constraints and a methodology containing epistemically sanctioned search and decision operations. These include stipulations of the form a law should take, viz. a statement of delimited universality with counterfactual import. Also required for the formulation of some problems is a language or conceptual system C that imposes constraints upon appropriate problems, including the machine searchable formulations under consideration, and an R specifying possibly effective empirical search and decision operations. With these givens, laws L or theories T are sought. The implementations of BACON under these specifications have reproduced 'discoveries' of Kepler's Third Law, Ohm's Law, Boyle's Law, the Boyle-Charles Law and others from given 'data' sets, where the data are specifications of actual empirical information E presupposing appropriate evidential relations R. In at least one case, viz. BACON 3's discovery of Kepler's Third Law, the presupposed R includes procedures for computing variables over which the search is conducted, viz. heliocentric distances and longitudes, from variables that can be observed, presumably celestial longitudes observed successively at the same time of year over a period of years. (Kuhn, 1957, gives a simple method for these computations for both 'inner' and 'outer' planets.) A parallel point can be made regarding Ohm's use of Ampere's instrument for measuring electric current.

As to the 'content' for representing problems, problem spaces and laws, the BACON programs generally contain designators or 'nominal terms', e.g. planet A,B,C, terms representing variables and various functions thereof periods, distances, $(\text{distances})^2$, periods/distances, and conditional rules, heuristics, for operating on these variables, e.g. 'If the values of a term are constant, then infer that the term always has that value', 'If the values of two numerical terms increase together, then consider their ratio'. (Langley, Simon, *et. al.*, 1987, Chap. 3). Also the terms have rules of usage associated with them, e.g. If A is a nominal term and A occurs in two contexts C(A) and C′(A), e.g. two rows of a table, then A=A, that is, A refers to the same object in each. If we are naming substances rather than individuals, then A=A will have to be interpreted differently. In the case of substances C(A) and C′(A) will refer either to the same sample of a substance or to different samples of the same substance. It should be clear that semantic considerations enter at least into our reading of the symbols that are written into the computer programs or that appear on printouts from the computer. As to other rules for terms, if P is a quantitative term and A is a nominal term, for every A, P(A)=p, that is, P has a definite value for every A, where p is a number or a vector, possibly natural, rational or real depending upon the application under consideration. Here again semantics must enter, for in some cases each individual is to be assigned a definite value for P and in others it might be that each substance is assigned such values. Thus mass would fit the first variant of this rule and density the second. This rule also plays a role in BACON.1's gathering of data, to which these questions are addressed (Langley, *et. al.*, 1987, p. 66): When the planet is A, what is P? When the planet is B, what is P? What is D? The first of these questions presupposes the above rule, viz. every planet has a period. In some applications for every A, P(A,T)=p, where 'T' designates a time interval. Such terms are appropriate for situations in which the value of P may change in time or on different occasions be different. Here occasions in T may be designated by nominal terms so that a row on a table might give data for the same object on different occasions. Also the same identity rule would apply to T that applied to A earlier. An example of such a quantity would be the planets' heliocentric longitude, which is time-dependent in Copernican theory. Correspondingly different questions will be asked: When the planet is A on occasion T, what is P? (Kleiner, 1970) These points further demonstrate that the BACON programs must have significant subject-matter specific preconceptions.

In some problems, e.g. those in which only one quantitative variable is specified for a number of specific named objects or systems, to obtain laws certain 'intrinsic' quantitative terms must be introduced. These include is quantities such as resistance, mass, refractive index, specific heat. Like other quantities, these also have associated rules which are presupposed in utilizing these terms for the discovery of laws. In the case of batteries and wires, it is assumed that the newly introduced terms, conductance for the wires and voltage for the batteries, have constant values for each wire and battery, respectively. Thus a rule for intrinsic terms is this: If X is a wire, then for all occurrences of X in this investigation, C(X)=c, where c is a constant whose value is initially computed from the currents through X in an initial step. In classical mechanics 'mass' would be considered an intrinsic term for each individual 'body' or 'particle'. Similarly: If B is a battery, then for all occurrences of B in this investigation, voltage V(B)=v, a constant whose initial value may be computed from the current through a wire of known conductance. Analogous rules can be given for the masses or specific heats of individual bodies, the specific heats of individual substances, optical densities of individual bodies or substances, etc. These rules are taken from preconceptions in the respective fields of electricity, mechanics, calorimetry, and optics. Again, the terms and rules are subject matter specific, but there are similarities among them, e.g. the rule assigning each wire a constant conductance is similar to the rule assigning each material body a constant mass. Both rules have similar heuristic roles in discovering respectively Ohm's Law and the Law of Inertia. By means of such similarities the 'intrinsic terms' can be usefully classified together as a *kind* of term for heuristic purposes. This classification gives rise to heuristically useful analogies which the proponents of BACON.4 exploited in its applications to hydrostatics, optics, calorimetry, and mechanics.

More specifically formulated, the heuristics for postulating intrinsic terms in the search for laws include these: If you want to *find laws* and you have observed a set of nominal terms A,B,...,X, Y,..., then *consider intrinsic properties* for A,B,...,X,Y,.... If you want to *consider intrinsic properties* for A,B,...,X,Y,... and I is a dependent quantitative term, and there is no intrinsic property associated with A,B,...,X,Y,... based on I, and the observed numerical values of I for observed configurations AX and AY are I_1, I_2,..., then *postulate* the intrinsic property C attributed to X,Y,..., *record* the values of C(X), C(Y),... as I_1,I_2,... and *store* them under condition A. (Langley, *et. al.* 1987, p. 135, paraphrased with some simplifications.) In short, the postulation of intrinsic terms is a means to finding laws governing objects for which we might already have a number of measurements of other terms, e.g. ammeter readings in battery-wire ammeter configurations, accelerations of objects under elastic force, thermometer readings for various samples of various substances, or angles of incidence or refraction for light rays entering and emerging from optical media. A historical parallel for this mechanizable heuristic for introducing concepts occurs when Galilean dynamics, which is confined to bodies acting under inertial forces such as gravity, is transcended. This occurred in the 17th century when mechanicians, such as Descartes and Huygens, address problems about the motions of bodies under other forces, such as impulse from collision or elastic forces. Under a given inertial force bodies of all weights, densities and magnitudes accelerate at the same rate, so no intrinsic attribute is needed to account for different accelerations under these circumstances. However, in collision bodies of different weights, densities, etc. can accelerate differently in response to the same impulse. Assigning a distinct mass-value that remains constant in time to each of these bodies will account for their different accelerations, and the accelerations, as Mach holds, can be used to measure these masses in initial experiments.

Constants, that is ratios, powers, ratios of powers, etc. of the 'data' whose values are the same for all data sets presented in the problem are found in BACON's operations, and Langley and Simon claim that these conceptual 'discoveries' both represent laws of nature and are not qualitatively distinct from higher-order theoretical conceptualizations. Other more 'conceptual' discoveries include the introduction of 'intrinsic' quantities such as voltage and resistance as determinable descriptions of objects known nominally, such as wire X, battery A, etc. In these cases the problem searched is more appropriately described as one in which metascientific, ontological and empirical beliefs M,B,R,E, and certain heuristics are givens and concepts C and laws L are sought. These problems illustrate the point that the boundaries of Kuhnian 'normal science' cannot realistically be defined very sharply, however, and the WSP's accessible to BACON are not strictly limited to Kuhnian paradigmatic 'routines', which presumably are puzzle solving routines using existing concepts and thus exclude conceptualization. In sum, there is a sense in which the BACON programs introduce new concepts, viz. the constant ratios and intrinsic properties which represent lawlike behavior.

As suggested by its name, the family of BACON programs is alleged by its authors and proponents to minimize the use of prior knowledge of a specific subject matter. They claim that the BACON programs make no special assumptions about the structure of the empirical 'data' from which they operate or the semantic meaning of that data. We have already seen some reason to doubt this point because semantic meanings are implicit in the background knowledge of the authors of the programs and occur explicitly in the programs in the form of rules for nominal and intrinsic terms.

Also many of the important heuristics in BACON are said to be 'data driven' in the sense that the conditions for applying the heuristics refer to some relations among the data. However, this last point is rather vague because of the vagueness of 'data', which for Langley, Simon, *et. al.* includes values of functions of primitive dependent variables, including those functions whose values are constants. The attribution of a constant to a function can be a statement of a law such as Kepler's third law, D^3/P^2=constant. In this last case the constant function can be considered a 'theoretical term', that is, a quantity that plays an essential role in a theoretically important law and is not directly observable, but whose value can be inferred from values of 'more directly observable' terms. Momentum and kinetic energy are terms of this kind. Special names for these terms in scientific practice indicate their importance due to their role in the more broadly applicable conservation laws for momentum and energy rather than any essential difference in the epistemic status of these and Kepler's term. Hence 'theory' and 'data driven' heuristics and research programs have been falsely dichotomized and at best indicate a difference in emphasis in programs that contain both heuristics whose conditions mention 'data' and those that mention higher order 'theoretical constructs'. Also Kepler's 'data' for the third law is laden with Copernican theory in the sense that neither heliocentric distance D nor the heliocentric period P of a planet is directly observable. It is a Copernican assumption reenforced by Kepler's celestial dynamics (see Kleiner, 1983) that these, among an indefinite number of variables attributable to planets such as synodic period, length of retrograde arc, latitude, are the most important or relatively essential variables for describing and explaining planetary motion. Also these quantities, unlike mean sidereal period, synodic period, celestial longitude and latitude, altitude and azimuth, are not directly observable, but must be computed using the Copernican configuration, as pointed out earlier.

It must be kept in mind that BACON addresses and solves problems in which a number of substantive scientific beliefs and peculiarly scientific values, not just logical and mathematical truths, are assumed in their formulation. These assumptions include the existence of yet undetermined intrinsic properties for a given subject matter. There is no *a priori* necessity that individuals be uniquely described by or have single valued attributes. Nor is there any *a priori* necessity that substances have such attributes. The desirability of obtaining constant quantities or the equivalent of lawlike uniformities is not intrinsic to human nature, but was probably invented by someone, perhaps a Pythagorean or perhaps Plato. Various kinds of symmetry, e.g. between juxtaposed material objects in gravitational or other mechanical experiments or between optical media of various densities, or the more abstract symmetries associated with conservation laws are equally contingent. The desiderata addressed by the BACON programs, viz. the search for laws and intrinsic properties, should not be philosophically controversial if it is recognized that these desiderata are only some among the many epistemic and other objectives of scientific research. BACON aids in the discovery of quantities whose values remain constant, i.e. the discovery of relatively low-level uniformities or laws, from given sets of values of various variables presupposed to be significant in a domain. The assumed 'givens' in this kind of problem can themselves be products of systematic inquiry that can differ significantly from the various heuristics exemplified in the several versions of BACON, and in various ways they may be sources of further inquiry.

However, our means of knowing the various 'givens' to which BACON is applied is yet obscure and is an item for further metascientific and meta-metascientific inquiry, or possibly even historical inquiry. Perhaps not in the case of the applications of BACON that Langley and Simon cite in their (1987), but surely in cases in which the empirical data, software and hardware is less available or more expensive or the inquiry process more time consuming and the subject less well-known, one must have some degree of prior assurance that the programmed search will be successful to warrant the initial investment required for its implementation. How do we know that there exist intrinsic properties for wires and batteries before BACON is put to work on their configurations and the ammeter readings produced in these configurations? What kind of knowledge of this subject is required to establish that the search for intrinsic properties is an appropriate scientific enterprise? This is the first horn of the dilemma Plato presented in the *Meno*: How can we, in a state of ignorance, define a subject-matter and a goal for inquiry? (Nickles, 1981, has raised these questions in a more general context; see below.)

Kuhn claims (1970) that the paradigm provides at least the information (part of the background B) needed to define a subject toward which such a search is directed. He also says that the paradigm provides the standards by which the successful completion of such a search is appraised (as prescribed by metascientific aims M in conjunction with received concepts C). However, the informational content of Kuhnian paradigms is notoriously, though perhaps in some respects properly, vague. One might suppose in this case that the Voltaic and Amperian 'paradigms' include conceptions of certain intrinsic properties of wires and batteries, e.g. electromotive force, voltage, conductance or resistance of wires and batteries, and that normal science under these conceptions seeks the quantitative evaluation of these properties for particular configurations of these objects. However, in this case BACON's search for intrinsic properties is a search for what is already known, and thus we are caught on the other horn of Plato's dilemma: There is no need to search for what we already know.

On the other hand, the supposition that intrinsic properties exist for wires and batteries might be plausible on historical grounds. History tells us that similar search programs for intrinsic properties of material bodies (e.g. mass) and substances (e.g. densities, specific heats and optical densities) have been successful and have produced lawlike regularities. However, the subject matters of these several research programs differ, whereas a Kuhnian paradigm is normally thought of as addressed to a single subject matter.

Perhaps the prior knowledge required for the application of BACON to wires, batteries, etc. consists of some general conception of intrinsic property as successfully applied to other subject matters and some warranted hope that this conception can be extended to this subject. Knowledge of such property types might be expressed in the form of the rule for intrinsic terms mentioned earlier. This conception of prior knowledge suggests an extension of the ordinary conception of a Kuhnian paradigm so that it covers more than one subject-matter or field of inquiry and makes claims about kinds of properties and perhaps even kinds of theory (Darden, 1982, Darden and Cain, 1989). This extension coheres with the generality of the BACON heuristics. None of these heuristics are restricted to one subject matter, but can be applied to quantitative data concerning the state and behavior of planets, physical and chemical substances, chemical elements and compounds, wires, batteries and electric currents, all of which are plausibly distinct subject-matters for scientific inquiry. What we are contemplating here is the possibility (and the plausibility) that there are 'interfield' heuristics whose application may be warranted by interfield metascientific beliefs M. This is not to say that the heuristics can be expected to be a global methodological principles for science because, although their scope may span several fields, they are still limited to certain fields and even within the fields in which they have been successfully employed.

Indeed the heuristics of the BACON program make no special semantic references to planets but can be applied to various lists of nominal terms and values for 'data'. They may or may not succeed in defining quantities that are constant, that is in finding laws in lists of data for dependent variables. This kind of subject independence is the basis of the strongest claim that can be made that BACON is a 'logic' of discovery, and beyond this, the heuristics share almost no distinctive features of sentential or first order deductive logic. They are neither universal, apodictic nor categorically normative, as is often claimed for genuinely logical principles.

Still, in any application of *bacon* one must choose variables which, in view of prior conceptions of the subject matter, are likely to yield laws. Indeed Kepler's term D^3/P^2 is not a concept that one would immediately think of in essentially characterizing observable planets, as would be the synodic period (period between successive locations relative to the Sun), which is directly observable and unique for each known solar planet. Still the former term would be chosen as a property of planets for the expression of Kepler's third law as an invariant for solar planets and over time. On the other hand, the choice of heliocentric periods and distances as descriptive terms is both presupposed in the usual expression of Kepler's law and essential to it, but this choice does not beg the question that Kepler sought to answer in seeking relations among these variables. That is, the choice is not made because the variables are essential to the expression of Kepler's law. Still the variables provide a simple approximative characterization of Copernican planets in the sense that every planet in a Copernican system must have a heliocentric distance and period if it is to be considered a planet, as opposed to a moon, star or the Sun. Of course it must be remembered that Copernicus' exact formulation of his theory did not use heliocentric distances or periods, but rather eccentric distances and periods determined with respect to the center of the Earth's orbit. Hence we have claimed that heliocentric periods and distances are actually Keplerian concepts because Kepler believed that the Sun moved the planets. For Kepler the power of the Sun was to be calculated by reference to heliocentric distance and its effects are manifest in the planets' periods. Hence Kepler had good reason to value as important the empirical law that relates the two variables, though his interest in celestial harmonies is what led him to its discovery (another example of serendipitous discovery).

A second strategy for the utilization of mechanizable search and decision procedures is the division of scientific problems into components, some of which are suitable for the application of mechanized procedures.

AI thus may provide heuristics for the solution of parts of scientific problems without being sufficient for solving the whole problem or all scientific problems: The empirical laws discovered by BACON, such as Kepler's law is only a partial solution to the problem of understanding planetary motion. BACON can also introduce a mass concept for data involving colliding bodies, and this can be construed as contributing to the problem of the planets. However, Copernicus' and Galileo's defense of a heliocentric system and even Newton's 'inductive' inference of celestial gravity (see Chapter 6 for a discussion of this kind of induction) seem clearly beyond the scope of the BACON programs discussed here or any plausible extensions of them. Similarly, the use of DENDRAL in the inference of plausible molecular configurations from mass spectroscopy is sufficient solving subproblems in molecular biology and MYCIN, for subproblems in medical practice (Feigenbaum, *et. al.*, 1971, Buchanan, 1982, 1985, Schaffner, 1985). Thesis IA also permits interfield borrowings of heuristics in the sense that there may be found in science procedures that are analogous, but not identical to those proven successful in the better defined contexts studied by AI and possibly those studied by cognitive psychology. For example, the various 'inductive' procedures, in a broad sense of any procedure of inferring or generating theory from observation, and 'deductive' procedures of elaborating testable consequences in science are analogous respectively to procedures in the BACON and PI (see below) programs and to the 'generate and test' heuristics (Newell and Simon, 1972) found in the various earlier discovery programs in AI. Paths of search in a problem space may be usefully analogous to the 'paths to discovery' described by historians, and other such analogies might be sought for more powerful or more specific scientific heuristics of more restricted scope (Simon, 1977, Kleiner, 1983). Of particular interest will be the 'generators' of successive cognitive states that might be apparent in various aspects of scientific research.

The development and successful implementation of programs like BACON is philosophically remarkable insofar as it clarifies by exemplification what heuristics can be and exemplifies the power of mechanizable heuristics under certain restricted circumstances for achieving limited and well-defined objectives. These programs also demonstrate the error in thinking that broadly 'inductive' procedures cannot generate conceptual novelty. They are also not limited to discovering concepts that are explicitly definable (N.B. it is doubtful that there is a distinct metascientific conception of 'explicit definition' from other forms of 'implicit definition') in terms of the descriptive variables used in the data. The intrinsic quantities 'gravitational mass', 'inertial mass', 'optical density', 'specific heat', etc. cannot be equated with 'logical constructions' from the kinematics of juxtaposed bodies, ray geometry, and descriptions of coincidences on scales without distortion, artificiality and inadequacy to these concepts in their actual depth. Presumably similar applications of one of the BACON programs will produce paradigmatic disposition concepts such as 'water-solubility', a concept which is comparable to density, conductivity, etc. in its applicability to chemical substances and its epistemic status of being not overtly manifest to human senses. Perhaps these or further variants of BACON can generate 'higher-level' theoretical concepts, e.g. concepts of internal energy and entropy, electric and magnetic fields, or even of properties for fundamental particles. Also the limitations now known for BACON do not imply that there are not other heuristics, other problem formulations, and other computational architectures which could be mechanized for generating these or other concepts.

5.3. PROCESS OF INDUCTION (PI)

Paul Thagard and several others have developed a program called PROCESS OF INDUCTION (PI) whose overall strategy is to appeal to whatever *relevant* prior knowledge is available in the solution of problems, where prior knowledge consists of an indefinitely large and ever changing fund of laws and facts. PI contains heuristics that use a richer fund of background knowledge, including pre-programed concepts C and laws L for specific scientific fields, than found in any of the BACON programs. Also PI learns from its own research experience: Successes in achieving sub-goals and goals, including novel and revised concepts, problems and records of procedures in solving them, are recorded in long term memory for future use, and these recordings can modify the fund of background knowledge. This feature embodies the important point that scientists and other investigators learn their craft from experience drawn from prior investigations (see remarks above regarding metascience and Chapter 6 below).

As in BACON, the laws L are expressed in the form of condition-action rules, that is, they have the general form 'If condition X holds, then do Y', where 'doing Y' might be activating another concept with its associated rules, as in inferring one transfers consideration or credence from one proposition to another. Laws thus have a practical component in that they can be used in planning and executing actions and they can be used inferentially. A law can tell us this: Conditions X and Y are sufficient for achieving Z. Thus if the production of Z is sought, according to this law it can be obtained by doing Y in conditions X. Alternatively, if knowledge of Z is sought, it can be inferred from knowledge of X and Y, as in the commonly understood form of scientific prediction and postdiction.

Its authors (Holland, *et. al.*, 1986, p. 3) address PI to these objectives, which should be compared to discussions the problem of scientific inquiry in earlier chapters:

(i) Design an artificial cognitive system that can process environmental input (observations) and stored knowledge (background knowledge in science) so as to benefit from experience in the sense that it is better able to achieve objectives of its actions. These objectives include the solution of epistemic problems, such as those of categorization and explanation. It has already been pointed out that it is useful to include epistemic objectives, the attainment of knowledge, among the other 'pragmatic' objectives of human activity, particularly when we are considering the activity of scientific inquiry. Such objectives, we have claimed, can be posed in the form of certain kinds of problem, epistemic problems.

(ii) Find procedures whereby a system can organize its experience so that it has some basis for action, i.e. some ground for presuming that action will achieve the system's goals, or is more likely with than without a certain organization to achieve the goals, even in unfamiliar situations. This objective is to enable a system to choose and implement appropriate heuristics, whether practical or epistemic.

(iii) Enable a system to determine that certain rules in its knowledge base are inadequate for achieving these objectives, and thus to reject them in efforts to solve the problem at hand. This objective is to design a system that can decide to abandon rules, concepts and heuristics that are non-performing and thus may be inappropriate or unpromising in a particular circumstance.

(iv) Enable a system to generate plausible rules, that is rules worth trying, and more adequate new rules, that is rules that are or promise to be successful in facilitating or achieving goals, to replace the inadequate ones. This objective includes refining rules that are useful but non-optimal, that is, rules that are successful in achieving some goals but for various reasons not completely satisfactory. The system should be able to respond to inappropriate heuristic choices with either others that are more appropriate drawn from the fund of background knowledge or by revising existing or creating new heuristics. These objectives include the search for metaheuristics, which may be construed broadly as including representations of knowledge and procedures for processing that knowledge, where each is formulated and adjusted to facilitate the other.

It would seem that any artificial model of inquiry that progresses in addressing these objectives would contribute to the philosophical problem of scientific discovery, for these objectives match nearly one-for-one the various features of this problem ascribed in earlier chapters.

The central problem of induction as seen by the designers and advocates of PI (Holland, *et. al.* 1986, p. 5, Thagard, 1988) is to specify rules of inference and processing constraints that will ensure that the inference drawn by a cognitive system (or a human inquirer) will be plausible, that is epistemically valued, or in plain language, believable, and relevant to achieving its (her) goals, that is, pragmatically valued, or, in plain language, useful. A philosophical assumption behind PI is that acceptable inductions are determined primarily by reference to current substantive knowledge in the inquiring system, not so much in reference to abstract logical or mathematical principles, but particularly knowledge about the variability of particular categories or domains. We shall call this philosophical position *naturalistic inductivism.* Knowledge of variability in categorization is needed in biological taxonomy, as pointed out earlier in the discussion of Darwin's biogeographical researches. Such current knowledge is largely subject-matter specific and less 'mathematical' in the sense of universal, abstract and apodictic (Holland, *et. al.*, 1986,

p. 22). 'Logical' principles of induction that generally prescribe such things as 'The more A is observed with B, the greater the probability that All A are B', or Bayes' theorem, are prominently absent in this program. Inductive generalizations are warranted just where substantive knowledge specifies a kind of uniformity like the following: We may inductively infer that terrestrial gravity applies to Jupiter's satellites because we have, during the period of the Copernican Revolution (roughly from 1543, the publication of Copernicus *De Revolutionibus*, to 1686, the publication of Newton's *Principia*), rejected Aristotle's 'lunar dichotomy', according to which terrestrial substances and laws are entirely distinct from celestial substances and laws, and its various implications. With Newton we assume that terrestrial and celestial matter is the same and that physical laws apply equally to terrestrial projectiles and the Moon and to Jupiter and its satellites. However, in our present state of knowledge we would also be cautious in drawing inductions from these objects to satellites close to the Sun or to black holes.

Thus induction is highly context dependent, guided by prior knowledge activated in particular situations confronting a system or inquirer as it seeks goals. Traditional philosophers of induction would question how knowledge of the relevant homogeneities and heterogeneities of the Universe is obtained for these inductions without begging questions about which inductions are permitted. Seemingly a defender of naturalistic inductivism would respond to this problem by noting that similar problems, e.g. the explanation of apparent centripetal accelerations near large masses, by default of indications to the contrary, give rise to attempted solutions that are most like solutions already obtained to these problems. In the absence information to the contrary, what other guidelines could be more reasonable to try? Furthermore, in the Copernican revolution evidence against Aristotle's lunar dichotomy, evidence that the heavens are like the terrestrial region, accumulated rapidly in the 17th century: From the very beginning it was never doubted that celestial and terrestrial optics were the same. This assumption is implicit in, e.g. early empirical arguments for the sphericity of the earth and the heavens. Galileo's use of the telescope depended upon this assumption, and it revealed geological features on the moon and meteorological happenings near the Sun's surface, viz. sunspots, both of which are evidence against the lunar dichotomy. The terrestrial impetus or inertia needed to account for the possibility of the Earth's motion can also maintain planetary motions in their orbits. Thus, given the optical homogeneity of the Universe, evidence accumulated that the Universe is also geologically, meteorologically and mechanically homogeneous in the sense that the same laws govern these phenomena everywhere. The initial assumption about optical homogeneity can be defended on the same grounds suggested earlier: In the absence of reasons to the contrary, the best bet is to extend principles already familiar in the terrestrial region to unfamiliar domains, e.g. the celestial domain. Why in such circumstances would one want to formulate and try unfamiliar principles, and what reason would there be, in such circumstances, to choose any particular one of these latter principles? (See Chapter 6 for further philosophical discussion of this kind of 'inductive heuristic'.)

Thagard and his associates thus view induction as a process of inquiry that draws upon prior knowledge of a specific domain judged relevant to a given domain specific problem, where that knowledge is a product of past experiences and actions, particularly efforts at problem solution, within that environment. We have already seen that problems contain informational and conceptual content: 'What is the mass of the Sun?' presupposes that the Sun has mass and all of the conceptual baggage that goes with 'mass' and 'the Sun', including post newtonian cosmology, the law of inertia, assumptions that the mass of a body is single-valued and relatively constant in time, that mass can be manifest in measurements of acceleration, gravitational accelerative force, etc. Posing this question activates our thinking in terms of all of these other concepts and possibilities associated with mass, among which various methods of measuring mass are particularly important for solving this problem.

In their efforts to simulate human expertise Thagard *et. al.* assume that expert inquiry is critically dependent upon specialized representations of knowledge about the relevant domain and methods that depend upon these representations. As such their approach opposes that of Simon *et. al.*, who seek, though do not really find, a domain independent 'logic' of inquiry. Furthermore, the Thagard group's strategy for representing knowledge is far closer to the analysis of scientific knowledge defended in earlier chapters than what is found in the BACON programs. For example in this analysis scientific laws and the concepts in them are inextrably linked together in determining the meaning, content or identity of both, and scientific laws play a central role in empirical methods in the form of subject-specific evidential relations. This conception of laws and evidence fits the representation of knowledge in PI by means of a multiply linked default hierarchy and network of rules, where nodes in the network are concepts. Hence we should expect as much if not a great deal more enlightenment regarding scientific inquiry from PI as from BACON. Particularly, from the plausibility of Thagard's claims regarding scientific expertise, we might hope for better models of scientific expertise whose heuristic power can be conveniently, rapidly and severely tested by running computer programs embodying those models.

For Thagard *et. al.* solving problems a process of retrieving an appropriate problem schema and providing it with problem specific parameters that presumably are given by the operator. The problem schema, in agreement with our own analysis of problems (see the problems Simon attributes to Keplerian inquiry above and also the discussion to follow below) has subject specific conceptual content. Thus the problem with its variable parameters specified provides information about concepts and specialized solution methods applicable in the circumstances presupposed by the problem. The problem posed provides goals and activates initial conditions within this network, and the activation 'spreads' to neighboring concepts or other relevant concepts by means of priority rankings among the rules. In the initial cycle most recently activated concepts in the relevant domain are tried, and if they fail to satisfy goal conditions a second tier of less relevant concepts are activated and tried, and so on until a solution is reached. This specificity is necessary for the activation of only that knowledge *relevant* to the problem at hand, and is thus an essential component of the overall strategy for finding practicable methods for solving the problems, methods that do not overwhelm the computer or researcher with exponentially exploding possibilities. Problem solving in PI is thus based in an essential way on domain-specific knowledge when it is available, though this is not to exclude using methods of greater generality like those used by BACON.

The system is also capable of trying alternative categorizations of the entities in the problem, sequentially by spreading activation, but even simultaneously along parallel paths. It can also generate novelty by combining concepts and seeking resolution of possible conflicts among expectations associated with each concept, as in 'feminist bank teller' (Holland, *et. al.*, 1986, p. 13).

Concepts are represented in this system as nodes in a network of rules, where the rules provide various kinds of relationships of mutual relevance among concepts. The rules hold by 'default' or defeasible presumption, and these defaults can be weighted in priority with the least central rule being the first to default. A default rule is to hold unless it is contradicted by more specific information. Thus birds are by default expected to be able to fly, but that bird X is a penguin will counter this expectation.

The following illustrates the basic unit of knowledge representation in PI, the *schema* (sometimes also called a 'frame', 'script', or 'concept'):

> Name: Sound
> Data-type: Concept
> Activation: 0
> Superordinates: Physical phenomenon, sensation
> Subordinates: Voice, music, whistle, animal sounds
> Instances: statement, shout, concert, bark, meow
> Activated by:
>
> Rules:
>
> Rule 0: If X is heard, then X is a sound.
> Rule 1: If X is a sound, then X is transmitted by air.
> Rule 2: If X is a sound, then X echoes.

Rule 3: If X is a sound and Y is a person and X is near Y, then Y hears X.
Rule 4: If X is a sound, then X spreads spherically.
Rule 5: If X is a sound, then X is a sensation.
Rule 6: If X is a sound, then X is a physical phenomenon.

The rules associated with the concept sound provide two kinds of links with other concepts that reflect features of the environment, (i) diachronic, as Rule 4 describes a process in time, and (ii) synchronic, as Rule 6 describes a relation of superordination (Holland, *et. al.* p. 15). The subordinate and superordinate slots and the rules link this concept with others and thus form an interconnected network that can be construed as a more specific representation of Quine's conceptual networks (1953). The default hierarchy fits an analysis of concepts that dates from Wittgenstein's analysis of 'game' (1953) according to which rules concerning language usage are not to be construed as necessary or sufficient conditions for application. These ideas have been extended by a number of writers to scientific concepts (Beckner, 1959, Ruse, 1977, Thagard, 1988).

Rules are represented as follows:

Name: Rule 3
Data-type: Rule
Concepts attached to: Sound
Conditions: (Sound ($X) true)
(Person ($Y) true)
(Near ($X,$Y) true)
Action: (Hears ($X,$Y) true)
Strength: 17
Activation: 0
Old-matches: nil
Current-match: nil
Satisfies goal?: nil
Projection status: nil
Current value: 0
Action instances: nil

The system also contains mechanisms for maintaining a set of 'facts', which are posted as 'messages' in the form '(Sound ($X) true)', where '$X' can be considered an 'open' variable that can be filled by a singular term, or can be an existentially or a universally quantified position. Posting this message is the equivalent of a person's affirming the proposition by making the statement "X is a sound". Such messages can appear in the conditions or the goal states of a problem, and as the problem is activated they become affirmations or posted messages.

The activation slot is the status of the concept, rule or problem in a given inquiry. Other slots for rules, such as 'matches', 'projection', 'status', 'action instances', 'strength', and 'satisfies goal' represent the record of the rule. A rule's strength determines its priority in the default hierarchy. Strength is a function of its past record of uses and its success and failure in contributing to problem solutions according to the formula: New strength = old strength − bid + payoff. Bids are subtracted from the rule's strength to keep it from repeatedly bidding and to reduce its relevance if it is not successful. Such loss in strength can be countered by a successful application, where even being a component of a successful application will generate payoff.

This method of appraisal can be compared to Lakatos' (1970) requirement that research programs be 'progressive' in the sense that they continually generate new observational contents. Stagnant research programs and rules that fail to succeed, though they may have no positive objections against them, are to be eventually set aside in both methodologies. In Thagard's system Lakatos' 'core', e.g. the general kinematic and dynamic assumptions and the gravity law underlying newtonian planetary dynamics, can be assigned a very high initial strength. Thus 'peripheral' beliefs, that is those lower on the default hierarchy, will be set aside in trials in the solution of, e.g., the problems presented by Mercury's or Uranus' anomalous orbits, in this case assumptions about the planets purturbing either orbit. But eventually the system will get to the 'core' and will attempt to discard newtonian gravity, newtonian inertia, the application of newtonian kinematical concepts to planets, or the kinematical concepts themselves, possibly in that order. As Lakatosians we cannot expect a solution to the problem when the search reaches the core unless replacements for newtonian gravity, inertia, or planetary kinematics are either available or can be generated by the system. At this level in science we are addressing conceptual problems, which we have described as the problem of finding a suitable system of laws and concepts for dealing with anomalous objects, events or processes that resist explanation with currently available concepts (Cf. Hanson, 1961, Chapter VI). As indicated above, Thagard *et. al.* have attempted to build the capability of investigating conceptual problems of this sort into PI.

The record of the action instances is of the messages that have activated it, and this record prevents the rule from firing repeatedly in the same cycle. 'Action instances' records actions of the rule with bound variables when its conditions are matched against posted messages. Thus facts about particular sounds can be recorded.

The 'satisfies goal' slot keeps track of whether the firing of a rule satisfies the goal of an active problem, that is, successfully solves it. 'Current values' are calculated when the conditions of a rule are matched, and this determines the bid a rule offers in satisfying activation conditions of further rules. The bid that an active rule offers for firing another depends upon its strength, the specificity of activating conditions and the support. The support for a rule is determined by the number of conditions activated and the number of times an activating condition is activated from different sources. When the strength of a rule diminishes to the point that it loses bids to rival rules it then is replaced, at which point the rival gains the opportunity to accumulate strength at the expense of the original rule.

Finally, the following schema represents a problem:

Name: explain sound
Data-type: problem
Start: (Sound ($X) true)
Goals: (Reflect ($X) true) (Propagate ($X) true)
Problem type: explanation

Activation: 1

Presumably the goal of finding laws and possibly a relevant concept linking some entity, process or event to sound propagation and reflection is implicit in the above problem statement. In other words, the machine or the person who conducts the search is programmed or informed to respond to the above statement with a search of the kind just described. Otherwise this statement of a problem would have to be regarded as insufficiently explicit specifying a goal state.

The start and goal conditions for an active problem such as the above post an initial message that activates rules for 'sound', 'reflect' and 'propagate'. Because of the network interconnections among concepts and the priorities in the default hierarchy, the activation of rules and concepts spreads horizontally to associated concepts and vertically to superordinates and subordinates, both to a limited extent. 'Support' for activating a given rule can accumulate from the activation and strength of the rules whose actions match its conditions (Holland, *et. al.*, 1986, p. 50). This cumulative support emulates the 'consilience' of inductions in categorization as several rules serving as criteria for applying a category become activated. Also activation spreads 'forward' from the starting conditions and 'backward' from the goal conditions of an activated problem. Furthermore, the activation of a number of rules at once simulates parallel processing.

Thus two further heuristics are incorporated into this system: (i) The 'backward' spread from goal conditions is like the method of analysis discussed in the previous chapter that prescribes that one work back from a conclusion seeking necessary conditions for a proof. (ii) The parallel processing emulates the heuristic of trying many possible paths of inquiry at once, a kind of 'hedging' whose effectiveness can be roughly estimated by multiplying estimates of the probabilities of failure for each path according to the multiplication rule for probabilities.

In hedging one wants to choose processes whose possible failures are independent. This point can be defended either by analogy or by reference to intrinsic features of the heuristic. One can argue from analogy with engineering strategies as follows (Wimsatt, 1981): One would not choose a backup braking system that works through the same hydraulic circuits that the primary system uses because hydraulic failures could cause both systems to fail. Rather one should design a backup system with separate hydraulics and other mechanisms each of whose functioning is mutually independent, for compound systems can break down if any one component fails. Hence the probability of breakdown of the whole braking system is the probability of failure of two independent systems, which is an intrinsic feature of a braking system composed of two or systems that function independently. The independence should extend to each component of each separate system, for an intrinsic feature of a compound system such as brakes is that the functioning of the whole system is as secure as the functioning of each component. That is, a braking system is a system in series, whereas when combined with a backup system the two function in parallel. Because the probability of failure in both braking systems should be considerably less than 0.5, the probability that both fail is exponentially reduced, to less than 0.25. This heuristic is analogous to the commonsense strategy of 'hedging one's bets', say by carrying a diversified portfolio of investments each of whose loss is hoped to be independent of other losses. However, losses in diversified investments are not always independent, for a recession or depression could bring them all down. Further hedging could be obtained by using investment strategies that profit from a falling market. In this case component strategies are chosen which are not mutually independent but whose probabilities of success or failure are inversely related. This second kind of hedging is different from the parallel hedging just discussed, but there are circumstances in scientific research in which an analogous hedging strategy could be chosen. Backup systems in an automated planetary explorer can thus be chosen or designed for various contingencies or unexpected changes in a planetary environment.

More specifically, parallelism in computer programming allows the system to take chances by activating rules on partially specified conditions and reducing the risk of failure by searching parallel paths as backups (Holland, *et. al.*, 1986, p. 24). However, in computer technology parallel processing at present requires a separate processor for each search path. Thus parallel search has the cost of more expensive hardware, and hence using parallel heuristics, like other heuristics, exacts costs as well as provides benefits (Wimsatt, 1980). In the organization of scientific research parallelism in a research program can be achieved by commitment of more people to the same project, particularly if such commitment also involves the use of a multiplicity of approaches, as in funding several approaches to the production of controlled fusion, (Thagard, 1988). The costs incurred in parallel organization are similar to those in programming: The more people or teams with distinct heuristics that are committed to a project, the greater the costs in terms of financial outlay or commitment of resources that could be utilized elsewhere.

Parallelism is thus a heuristic that has limited generality. It applies to research projects with common goals but distinct heuristics as well as in computer programs and it is promising in both cases for essentially the same reasons. The limits to its generality include the costs incurred: If a strategy has a high chance of success, then the expense of parallelism may be unwarranted. In the organization of scientific research the success of a research program may not be sufficiently important to warrant the expense of funding backup heuristics. A second possibility of failure is that in unforeseen circumstances the parallel strategies chosen may not all fail independently, a possibility whose knowledge might be drawn from the investment analogy cited above. In engineering the radiation effects of a nuclear blast could cripple all of a system of normally independent parallel communications strategies. Parallel research programs can be exploring domains in which it is far from a certainty that normally independent strategies will remain so in the domains to be explored.

(iii) A record of various rules' strength also provides a trouble shooting heuristic when multiple rules are used in a process of inquiry: First suspect and check out the weakest rule in this group. This rule will be the one that is the lowest in a default hierarchy. One can also appeal to commonsense in defending this heuristic: The experienced mechanic learns from past practice which components of the Jaguar are the weakest, and when symptoms of abnormal performance occurs she replaces them in order of the weakest first, while always operating within the constraints of possible causes of particular observed symptoms (Holland, *et. al.*, 1986, pp. 71f). In the process she may restore components that have been replaced if her diagnosis has yet to succeed in getting the car to run properly.

Activities corresponding to an inquiry in this production system cycle with 3 steps in conducting an inquiry: (1) match facts with rules to determine which conditions satisfy them and (2) a subset of the matched rules, the strongest or those with the highest bids, are selected to be fired. This procedure (2) is necessary for narrowing the paths of exploration to some of the more relevant concepts and rules, and thus 'spreading activation', a general feature of the architecture of the PI programs, is technically a heuristic. (3) The selected rules are fired to take their specified actions, including the posting of further facts to be matched in (1). Since the actions of rules are to post messages that generally differ from their action conditions, a different set of rules will typically have their conditions matched for the next cycle. Posted messages can provide bids for the firing of further rules and thus the support for a rule can be passed through localized regions of the conceptual network causing the spreading activation of rules.

Most interesting for the purposes of this book are the heuristics bearing upon the generation of novel rules and concepts. Some that are incorporated into PI are as follows:

(i) New rules may be applied in situations where none of the high strength rules, that is the relatively successful rules in the background knowledge, have their conditions satisfied. Such situations might typically be presented in novel problems, particularly problems concerning domains not dealt with by established rules. The distinctness of the domain should be represented in the problem's initiating conditions, which also serve as activating conditions for relevant rules and concepts. In these cases rules may be activated by partial satisfaction of their initiating conditions.

(ii) When failures of prediction occur in the application of a complex of rules and concepts, the strategy, as implicit in the default hierarchy conceptual structure, is to challenge the weakest rules first. This procedure is similar to Lakatos' 'negative' and 'positive' heuristics which protect a 'core' of assumptions common to a family of models or theories and recommend that 'auxiliary' assumptions peculiar to a model or common to a subset of the models be challenged and replaced in the effort to develop new models. However, contrary to Lakatos' position, there need not be a predetermined distinction of kind between 'core' and 'periphery' where the former are considered absolutely beyond critical assessment. Rather it seems truer to the history of science to think of these components as forming a continuum or differing only in the degree to which it is considered plausible to challenge them (Laudan, 1977, Chapter 3). PI carries this relativization of Lakatos' distinction a step further in proposing that the 'centrality' of a rule is a function of its past track record in solving problems and is always subject to revision as inquiry proceeds and further successes and failures are recorded. This last step is supported by the commonsense diagnostic logic used by the Jaguar mechanic. Rules can be replaced individually, though this happens after a rule looses strength relative to others by placing bids that do not succeed in contributing to problem solutions. Thus 'core' or strong rules are not as such immediately replaced, though they may be replaced by dropping out of the core, that is, by losing strength. Candidates for replacement rules are drawn from close associates, as determined by synchronic relations such as superordination and subordination, to rules that are already though decreasingly active.

(iii) Higher order generalizations, often generalizations across domains, and domain independent heuristics provide additional guidelines and constraints in the search for new rules and concepts: Among these are types of categories, e.g. classes, relations, transitive or intransitive relations, the quantities, vector, scalar, nominal terms and intrinsic properties for which BACON has characteristic rules. Category formation also can be guided by highly general rules about expected distribution of properties across a category. Thus the trained ornithologists that Darwin consulted had general knowledge of various kinds of variability within species of wild birds, which were useful in identifying the new species represented by Darwin's specimens. General causal schemes have associated with them general rules for reasoning about classes of causal relations, multiple versus single causality, deterministic versus probabilistic causality, processes that are influenced by internal tendencies and external relations, such as the orbiting of a planet around the Sun or the movement of a thermal system to equilibrium.

The Aristotelian causal scheme illustrates historically this kind of causal typology. It includes final causes or intrinsic teleology in addition to the material (composition), formal (type of system) and efficient (operating) causes attributed to these processes in contemporary physical science. Newtonian science and the Deistic programs that followed Newton retained material, formal and efficient causes in classifying and explaining the behavior of mechanical systems, but replaced Aristotle's intrinsic species-specific teleology with an extrinsic teleological agency that provided the design needed for the initial conditions of the solar system and the for the functional organization of plants and animals.

(iv) Heuristics of 'conceptual recombination' exploit combinatorial advantages of building blocks and hierarchies whereby a large variety of concept-rule structures can be made by selecting, reordering and rearranging common components. This heuristic, like (ii) above, assumes that rules and concepts can be moved about from one to another context as modular units. The explanatory heuristic behind atomic theory, where unity and economy is achieved through common components and diversity is achieved by different relations among the components, is similarly valued because it exploits these combinatorial advantages of multiplicity from few basic elements (Harre, 1972, Chapter 5). There are two sources for new rules: (a) Inputs from the environment, that is the outcome of observation and perception, can exhibit novel combinations of properties and relations, as might be found from inductive associations in scientific investigations, and (b) existing structures in the background knowledge can be transformed or combined, possibly as an outcome of non-inductive efforts at concept formation. Recombination can operate at different levels:

Parts of conditions and actions of individual rules can be recombined to form new rules, as can entire conditions and actions. If rules are organized into sets, as by rule couplings or in the formation of concepts, rule sets can be combined in various ways to generate new concepts. At least some of the modular units of rule-concept structures surely are such sets, and in modular separation and combination the identity of most of the rules and concepts included in the set can be maintained.

Conflicts between rules can be generated by combining previously distinct concepts, and can occur when a search simultaneously activates conflicting rules, an occurrence that seems particularly likely in systems that conduct research on two or more parallel paths. In PI such conflicts are to be resolved by inactivating defaults via priority ordering in rules, as represented by relative strengths.

The strategy of combining rules has a parallel in the history of quantum mechanics (Hanson, 1961, Chapter VI): Einstein's study of photoelectricity suggested that light energy, already successfully described as propagated by waves, can be discrete and spatiotemporally localized, as one might expect from the impulses produced by particle impacts. DeBroglie's thesis begins with classically known similarities in wave and particle propagation such as the possession of momentum and concludes by also attributing wave characteristics to the motion of a material particle such as an electron. However, this combination of wave and particle concepts initially seemed to produce conflicting conceptual content: Particles must be localized in space and time to be distinct as individuals and from another kind of physical entity, a field. They must possess what may be called precise Q-properties (in Hamiltonian mechanics position coordinates are designated by 'q') to be distinct or well-defined particles. On the other hand a wave cannot be identified as such unless it is conceived without the spatiotemporal localization needed to distinguish a particle, otherwise its wavelength or its period is indeterminate. In DeBroglies' wave mechanics definitive waves have exact momenta, that is, wavelengths and phase velocities, which may be called P-properties (after Hamiltonian 'p' for momentum).

Quantum theorists tried several strategies in attempts to mitigate the 'paradoxes' of this combination by one or another of several ways of forbidding the conjunction of P and Q properties in any one dimension. One strategy appears in Neils Bohr's 'complementarily' thesis: Wave characteristics are manifest only when certain kinds of measurement are carried out, e.g. diffraction and interference measurements, P-measurements, in which wavelength and momentum can be determined. Particle characteristics are manifest only when another kind of measurement, localization measurements or Q-measurements by effects of collision with ionizable or excitable molecules, are carried out. The physical conditions for P and Q observations or measurements are mutually exclusive. Hence as long as one considers only dispositional properties in quantum mechanics, e.g. the disposition to have a determinate direction of propagation without localization, or the disposition to be localized, without asking for some characteristic of the quantum object underlying these dispositions, the seemingly contradictory consequences of wave-particle duality can be avoided. In this case ontologies including fundamental particles are maintained but restrictions are imposed upon circumstances in which they are ascribed location or momentum, respectively mutually exclusive Q-conditions and P-conditions which are activated respectively by position and momentum measurements. Alternatively, one might claim with Bohr that the subject matter of quantum mechanics is not photons, electrons, or other fundamental particles, but rather wholes composed of quantum object and measuring apparatus. This is a shift from the usual quantum ontology. Localization, on the one hand, and momentum, on the other, are the usual classical concepts but each can only be attributed to one of two mutually exclusive types of unanalyzable object-instrument compounds, Q-types and P-types. These types are respectively position and momentum measuring apparatus in combination with the quantum object.

In either of these cases, as well as in the subjective interpretation of Born's view that the psi function represents a probability of localizing a quantum object, the belief that quantum objects can and should be described independently of conditions and apparatus of observation is set aside. This belief is a component of traditional scientific realism, according to which unobservable objects introduced for explanatory purposes can be ascribed certain properties independently of our observation of, knowledge of, or means of knowing them. This belief was very high in any 'default hierarchy' that can be attributed to classical physical concepts, particularly to the 'primary qualities' attributed to atoms or fundamental particles. The heuristics of PI would relegate the removal of this belief to stages that appear very late in the conceptual inquiries that follow Einstein's and DeBroglies' discoveries, a point that seems true to history. Schrodinger's efforts to dispense with the particle concept and to consider, for example, the electron as a vibrating field or charge 'smear' emphasized P-properties to the exclusion of Q-properties. His efforts ran into problems with 3N dimensional wave functions needed to describe N body quantum systems and with the instantaneous 'collapse' of an actual wave when it encounters a localization experimental setup. Hence this attempt to remove the duality is as objectionable as clinging exclusively to the particle picture in attempting to describe interference experiments, which would seem to require instantaneous action at a distance. Both attempts have been abandoned by quantum theorists.

However, once classical realism is questioned and rejected, the old concepts of wave and particle are brought back and applied with the qualifications of Bohr's complementarity. The reason for this is that quantum phenomena retain many of the features, such as interference and collision behavior, that have been successfully explained in classical theories by P-concepts and Q-concepts respectively. Bohr tells us that classical descriptions apply to object-instrument compounds, or they can appear in the descriptions of dispositions that cannot be further analyzed. In the terminology of the PI programs these concepts have a residual strength based upon past performances in classical physics, and this strength warrants their return after realism is weakened, even after they were unsuccessfully set aside in earlier stages of the inquiry. As the system conducts its search by initially setting aside higher level principles it restores relatively strong principles that are lower on the default scale.

Moving a single rule from one context to another raises questions about the identity of the concepts it incorporates, as some post-positivist philosophers have suggested (Feyerabend, 1962, Kuhn, 1970), and consequently questions are raised about the identity of the rule itself in a metascientific theory in which concepts are viewed as nodes in a network of rules. However, in the descriptions given of PI rules are identified by their activating and action clauses which contain terms which, though probably involved in other rules' clauses, have established meanings outside of the concept-rule network in PI. That is, PI does not seem to have built into it anything like a full set of semantic rules for terms such as 'person' or 'hears' in the rules for 'sound'. Thus it seems that the identity of what amount to certain prescientific concepts is presupposed in the representational system for PI, and it is in terms of these concepts that rules may be identified. Scientific concepts, then, are identified in terms of the set of rules that join in some node in the representational system. Moving individual rules from one context to another may thus generate new scientific concepts, but the prescientific concepts used to identify the rule need not be thus altered.

Rule coupling is basic in concept formation, where one of two coupled rules may contain in its consequent activating conditions for the second giving a temporal sequencing of the rules, or their activating conditions may be sufficiently similar that they become synchronically associated. Couplings can be generated in response to observed covariations and temporal relations, as prescribed by inductive procedures in science. The operating principles for coupling are triggered when strong uncoupled rules are active in the same contexts, as in the solution of a number of problems. In parallel systems several rules can be active in the same context, and these can then be associated as mutually relevant, possibly by adding common activating conditions or removing conflicting conditions.

(v) Nearly all inductive inferences consist of some form of generalization or specialization. Rules can be generalized by removing activating conditions and specialized by adding them, that is by de-restricting and restricting their application. This is one form of generalization called 'condition simplifying generalization' (Holland *et. al.*, p. 88). Category formation can involve either generalization or specialization depending upon what is done with the activating conditions for the category when rules are synchronically combined to form a node in the conceptual network. One form of analogy involves generalization by removing some of the activating conditions from rules or concepts. Thus the activating conditions for 'selection' in its pre-Darwinian usage includes 'domestic organisms'. In creating the concept of natural selection Darwin removed 'domestic' from these conditions thus rendering 'selection' applicable to all organisms, wild and domestic. Similarly certain activating conditions differentiate 'species' from 'variety', such as 'created by God' and 'produced naturally': By removing the former condition he merged 'variety' and 'species' into a continuum of heritable differentiation associated with a common cause.

Another form of generalization is called 'instance based generalization'. Here a new rule is produced by associating two or more features found in specific instances. This procedure takes into account the known variability of the kind of object in question in regard to the kind of property about which generalization is contemplated. For example, suppose the following rules are strong: If x is large-winged, brown, feathered then x can fly. If x is small-winged, brightly colored and feathered, then x can fly. Accordingly and in the light of general background knowledge that generally among insects, bats and birds wing size is not a deterrent to flying these rules may be replaced by: If x is feathered, x can fly. This heuristic can be considered an application of Mill's method of agreement according to which what can be eliminated among correlates is not a cause. Since wing size varies among flying animals, no particular wing size is causally necessary or sufficient for flying. It might also be used in taxonomy for the following change of rules: If x is large-beaked and about the size of a robin, arboreal in habit, has a pointed crest on its head, is all red with a black face, then x is a northern cardinal. If x is large-beaked and about the size of a robin, arboreal in habit, has a pointed crest on its head, is buff brown with a black face, then x is a northern cardinal. These two rules can, in the light of general knowledge of sexual dimorphism in birds, be replaced by the more general rule: If x is large-beaked and about the size of a robin, arboreal in habit, has a pointed crest on its head and is either buff brown or all red with a black face, then x is a northern cardinal. Here being red and being buff brown are eliminated as separately necessary for being a cardinal and replaced by a disjunction between the properties, a move which has the effect of generalizing the cardinal rule, though the rule still retains at least one of two colors as a necessary condition.

When predictions fail, specialization can be carried out by qualifying a rule. Qualification can be done by adding clause to the initiating conditions of a rule. Thus the properties of a penguin are not predicted by its classification as a bird or even as an aquatic bird. However, its unusual shape, habitat and locomotion, what amount to penguin specific properties, can be construed as predictors of probable flightlessness. Thus a modified default rule can be created from a parent default rule by augmenting the initiating conditions of the parent with the complement of these penguin specific properties in specializing the rule's activating conditions. If x has wings and is not pudgy and waddling with hunting habits like a seal, then x can fly. A second heuristic might be added at this point: If specifying properties are not known for the instance in which the prediction fails, activate a search for a specifying property that meets certain general and domain specific constraints, such as Simon's intrinsic properties. This strategy of specification has often been evident in the history of physics. Newton (1962, Vol. I, Book III, Prop. XIV) proposed the following rule: If x is a (known) planet and obeys Kepler's first law, then x's line of apsides is fixed in space or can be accounted for by purturbations from other planets. However, in the 1820's it was discovered that the line of apsides in Mercury's orbit precessed at a rate that cannot be accounted for by purturbations from known planets, such as Venus, Earth or Jupiter. Hence Newton's rule fails to predict. However it can still be applicable with a qualification: If x is a (known) planet other than Mercury then its orbit does not precess at a rate that cannot be accounted for by purturbations from other planets. Similar specializations are introduced for Newtonian rules of mechanics that excludes their application to quantum mechanical objects such as electrons, protons and photons (Holland, *et. al.* p. 89): If X is a body whose action (energy X time) is significantly larger than Planck's constant (6.55×10^{27} erg. sec.), then X can be specified by an exact position and velocity.

5.4. CONCLUSIONS

1. Machine inquiry is an aid but not a replacement for human inquiry. Certain subproblems of a research program, such as identifying a chemical or diagnosing a disease, or even finding a law or introducing a concept for finding a law for certain kinds of data, may be routinely solved by an appropriately programmed computer. However, we can hardly expect computers to conduct all aspects of scientific research or even emulate many of the historically significant scientific discoveries. Still mechanizable heuristic programs may be useful in suggesting, and even testing scientific heuristics. However a strategy for discovering and evaluating heuristics is to borrow like or analogous heuristics from commonsense, the history of science, and empirical psychological investigations of human problem solving practices, and to initially appraise them in terms of confidence in them established in these other fields.

2. WSP's for machines have decidable answers, but such decidability is generally not possible in science. There is no routine decision procedure for determining whether most scientific propositions are known, as must be done in answering epistemic questions in science. Science or a scientific discipline is not a deductively closed system based upon a definite list of unchangeable axioms and rules of inference, but consists of families of fallible and replaceable principles as well as problems and their presuppositions, and changeable values or standards of truth and scientific adequacy. Better scientific problems are those that presuppose more substantive information or further constrain, with justification in the light of current empirical information and epistemic standards,

possible solutions. The criteria for valuing problems cut across Kuhn's distinction of normal from extraordinary science. On the one hand conceptual problems appearing in extraordinary science can be well formed in the sense that conditions of adequacy for some sought after concept, e.g. that it solve some problem of explanation and meet some epistemic requirement, can be well-defined so that satisfaction can be routinely decided. On the other hand an epistemic problem in normal science, e.g. finding suitable evidence for the presence, at the center of the Milky Way, of a black hole conceived routinely by a general relativistic model, can be ill-stated in Simon's sense because we cannot routinely determine what will solve it. Part of the problem is to determine what kind of evidence will indicate the black hole's presence.

3. Routines for producing intrinsic variables falsify the notion there is no rational procedure for scientific discovery, particularly for novel conceptualization. These routines are not formal or algorithmic routines that are conceived *a priori*, but are based on procedures known *a posteriori* to be successful in other fields. Such knowledge constitutes a kind of trans-disciplinary paradigm. Intrinsic terms are a kind that do a certain job: They ascribe a property that is introduced to distinguish systems that behave differently. The heuristic that Simon *et. al.* use for introducing intrinsic terms can be extended from optics, thermostatics, electrodynamics and mechanics to nuclear physics, and can be informally stated thus: When a distinction must be made between objects, events or processes to get a law, introduce a descriptive term and initially evaluate it by the needed distinction. Then, since it is presumed constant for the entity to which it is ascribed, use it with this value in other contexts. This is a conceptual heuristic, i.e. a method of concept formation. Conceptual innovations thus are not necessarily irrational, uncontrolable and unguided mental processes that terminate in surprising, unplanned illuminations that we can do nothing to facilitate.

Kuhn's distinction between normal and extraordinary science is drawn in terms of routine and non-routine practice, as well as in terms of procedures for application of established concept versus unguided conceptual innovation. Here we see something like a routine for introducing concepts, though not a universal routine, but a heuristic. Thus Kuhn's description of the distinction is not quite correct: If extraordinary science that in which conceptual innovation occurs, extraordinary science can also be routinized at least in addressing some problems.

Other heuristics of concept formation include analogies based upon constraints from general conceptual categories, combining concepts and various procedures for adjudicating conflicts between them. The search for analogous concepts is activated by analogies in problems: The empirical observation that numbers of ejected electrons depend upon the intensity of the light hitting them and the energies depend on the frequency of that light activated in Einstein's mind the conception of electrons being ejected by collisions with light 'particles'. This interaction belongs to a category that includes billiard balls, molecules exerting pressure and transferring energy to other molecules, etc. The combination procedures might be activated by the empirical correlation of conditions under which each concept is normally applied or, as in the case of quantum mechanics, the theoretical discovery that waves and particles share properties such as momentum combined with experimental discoveries of wave localization. Several generalization strategies include extension of concepts to new domains, as Darwin extends selection from the production of varieties to species and Newton extends gravity from the terrestrial to the celestial region. In both cases an indicator that generalization might be appropriate is continuity, in Darwin's case between variety and species and in Newton's case between terrestrial projectile and satellite. A second generalization heuristic involves the elimination of conditions on the application of a concept, which can be implemented by an extension of Mill's method of agreement, or what amounts to Socratic counterexamples. Conceptualization may also occur by specialization, which is heuristically appropriate when a prediction fails. When Galilean dynamics fails to predict the behavior of impacting billiard balls, say by anticipating equal motions after the impact when the motions are actually unequal, the descriptions of each can be specialized by introducing an intrinsic variable, mass, and measuring it as the inverse ratio of motions after the impact. Thus the heuristic for introducing intrinsic variables employed in BACON can be construed as a special case of the specialization heuristic of PI.

These illustrations should establish that conceptualization in science can be subjected to a kind of 'logic' or 'rationalization'. Scientists researching problems of explanation do not have to just hope for the conceptual discoveries that bless with illumination the dedicated investigator who appropriately reposes. Rather a variety of possible procedures of analogy, combination, generalization and specialization are available, and each problem may be accompanied with indicators regarding which procedure has some likelihood of success.

4. Simon's BACON programs are not really 'data driven' research strategies, first because they are searches for laws, that is quantities that are functions of given values of independent variables that remain constant, where the nature of such laws are learned from other disciplines or subjects in which laws are already known. That is, they are driven by the search for a law, which is an objective not determined by 'data'. Secondly, the 'data' from which the research begins is 'theory laden' in the sense that it consists of values of variables that are of importance for theoretical reasons, or variables whose importance presupposes theoretical beliefs. Hence the limited success of the BACON programs supports no form of inductivism, the view that science is produced by routine inductive inferences without any preconceptions from theory free empirical 'data'. The objectives of research conducted by the BACON programs as well as the conception of the subject matter of each and the choice of variables for data all presuppose a background of factual, nomological and metascientific belief that can be broadly called 'theory' and 'metatheory', as is readily seen from an examination of the content and operations of the BACON programs. Contrary to Simon's claims, these programs are theory or metatheory driven and in no way challenge, but rather support the overall viewpoint defended in this book: Scientific inquiry must be conducted within a complex of ontological, nomological

and metascientific background beliefs that provide content, validity and importance to the objectives of inquiry, that is, to the problems or questions to be investigated.

5. The program PI of Thagard *et. al.* models human 'induction', which can be construed very broadly as categorization and explanatory inference as well as the solution of practical problems, problems of choosing what to do in order to achieve a given objective, whether epistemic or other. Induction is construed naturalistically in the sense that it is based upon, and justified by, substantive knowledge of the uniformity or homogeneity of a domain. Advances in inductively generated knowledge also produce further advances in the powers of inductive reasoning. This naturalistic construal is shown to be much more in conformity with scientific practice than are more traditional appeals to inductive 'logic' in its various forms: Traditional inductive logic includes *a priori* assumptions about states of the world and uniformities in the world, as in the case of Carnap's approach to the problem, or the mathematics of probability, as underlies 'principles of induction' and Baysean probabilities. These probabilistic approaches also make either questionable *a priori* assumptions about the probable uniformity of nature and questionable assumptions that epistemic valuations in science can be construed as probabilities.

In PI knowledge is represented as clusters of factual and nomological beliefs which center upon concepts and domains. This clustering provides the relevance relations needed to call forth from a general fund those items of knowledge most relevant to a given problem. The overall strategy for solving problems is to use the knowledge presupposed by a problem to call forth relevant facts and laws from background knowledge for the solution of the problem.

Records are kept of the successful use of a relevant principle in a given context, which strengthen its relevance in that context, where the context is a domain of inquiry specified by a problem's presuppositions and objectives. Conversely, if a relevant principle lies unused or fails to contribute to successful problem solutions, its relevance is diminished at the expense of other relevant principles. Thus principles can migrate from a 'core' of credibility and importance to a 'periphery', where the difference between the two is a matter of degree in a default hierarchy of rules connecting concepts.

Shifting relations of relevance can initiate the combination of concepts which may have conflicting rules associated with them. The heuristic for dealing with such conflicts is to drop the rule in a conflicting pair that has the least strength in the default hierarchy. Such a heuristic was used successfully in the conceptual development of quantum mechanics. PI also uses a heuristic that activates rules 'backward' from the goal conditions of a problem, a heuristic that is similar to the classical 'method of analysis' where one reasons from a theorem to be proven back to axioms and prior theorems that might contribute to its proof, in the light of conditions for what would be acceptable as a proof. Furthermore the parallel processing capability of PI is comparable to the heuristic of hedging investments by using a multiplicity of independent vehicles.

6. The above reflections illuminate several aspects of a methodology for heuristics, that is, some normative guidelines for their discovery and appraisal.

First of all, there are analogies among heuristics, e.g. the wisdom of carrying a spare tire and parallelism in inquiries, whether conducted by teams of humans or by several processors in a computer. The analogy alone is sufficient to establish the plausibility of the heuristic in a domain in which its efficacy is unknown. The wisdom of carrying a spare tire is well known, but one might argue for the wisdom of hedging with multiple investments or lines of inquiry just by pointing out that it is similar to carrying a spare tire. Such an argument would be appropriate if there were no known intrinsic characteristic of the procedure to warrant its heuristic merit. It can also reveal where the heuristic may break down, say in circumstances in which the parallel components do not function independently.

However, the heuristic merit of a procedure may also depend upon intrinsic attributes. Intrinsic features are ultimately important in deciding the worth of an argument from analogy in support of a heuristic. Like other items, there are many respects in which heuristics are analogous, and some analogies between heuristics whose value is known and those that are unknown can be entirely misleading if the common attributes are not what intrinsically makes the heuristic work. Thus hedging evidence with consilience might be compared to the redundancy of reading duplicate copies of a book if one is entirely ignorant of the probabilities involved in hedging with backup systems. In the case of hedging in a program several strategies are pursued at once in which each has a certain probability of failure. If the component strategies can succeed or fail independently of each other, as should be chosen in hedging, then the probability of the program failing is the product of the probability that each component fails. Since the probability of failure of any properly chosen component should be less than one, preferably less than half, parallel hedging should produce geometric reductions in probabilities of failure, a consequence that gives parallelism great heuristic merit.

CHAPTER SIX

THE HEURISTIC CHARACTER OF TRADITIONAL SCIENTIFIC METHODS

6.1. INTRODUCTION

The history of philosophy contains a number of attempts to formulate a procedure for generating theoretical knowledge, proofs, knowledge of causes, forms, essences, first principles and the like. These have been variously described and named as methods for inferring causes from effects, methods of analysis, induction, retroduction, and abduction. The methods described under these headings are often described in contrast to inferring from causes to effects, and methods of synthesis or deduction. The latter methods are plausibly understood as methods of applying established belief or principles in proving theorems or in explaining or predicting various occurrences, and philosophers, both recent and earlier, have viewed the latter methods as essentially deductive and philosophically unproblematic. However, the former methods are seldom sharply distinguished and seemingly several distinct procedures are designated by terms such as "analysis" and "induction". They have also been generally regarded as philosophically problematic, but still the hope of discovering such methods has risen repeatedly in the history of philosophy. (See Oldroyd, 1984, for a useful summary account of both kinds of method.)

Also scientific methodology has been construed as aiming at the elaboration of rules governing the conduct of all scientific research, that is, it is the search for global prescriptions for scientific procedure. Paul Feyerabend (1970, 1974) has subjected various such rules to attack, most notably Popperian rules of empirical falsifiability and non-*ad hocness*, and the logical empiricists' rule that novel theoretical developments be consistent with received background theories. Feyerabend argues that if scientists were to apply these rules rigidly and without qualification important discoveries, such as Copernicus' theory of the planets, could not have occurred. Copernican theory contradicted received cosmology, gravitation theory and dynamics, required the introduction of *ad hoc* dynamical principles such as circular inertia, and was contrary to received observations such as the absence of very strong tendencies of projectiles to move to the west.

What Feyerabend establishes is not what is suggested by his rhetoric, that scientists ought to embrace methodological anarchy or the contraries of received methodological principles, such as 'counterinductively' accepting only hypotheses that contradict experience. Rather, his argument is properly valued for showing that traditional methodological rules are appropriate only for certain epistemic circumstances. For example, 'inductive' rules requiring at least that we entertain only hypotheses compatible with observation are appropriate just in circumstances in which there is no current or reasonable suspicion of conclusions drawn from observation. Where there is reasonable doubt about the reliability of such conclusions one may or perhaps ought to proceed counterinductively. In the case of the Earth's rotation the Aristotelian principle that the horizontal motion of all bodies requires a sustaining force underlies the interpretation of the 'tower experiment' as ruling out the Earth's rotation. If such forces are required, objects dropped from a tower on a rotating Earth would reach the ground at a considerable distance to the west from the foot of the tower, which is contrary to commonplace observation. However the inadequacy in its account for projectile motion is reason to suspect the Aristotelian principle, and, as a consequence, there is also reason to suspect the traditional conclusions drawn from the tower experiment. In such circumstances one may at least consider the Earth's rotation as a possibility for further investigation.

In sum, Feyerabend's arguments do not establish that all methodological principles should be rejected for all scientific endeavors. When induction is rejected it might appropriately be replaced by some kind of counterinductive procedure. The rejection of induction under one set of circumstances does not rule out its reasonable and possibly successful use under other circumstances.

Another of Feyerabend's arguments against traditional methodology is that the various components of scientific belief may be variously developed, credible or reliable at various times in the history of a scientific discipline. These components include some discussed at length earlier, e.g. the evidential relations or auxiliary laws needed for designing, executing and interpreting experiments, the background beliefs that provide an ontological context for scientific inquiry, or the mathematics and other conceptualizations governing capacities for formulation, belief and deduction. This argument suggests possible conditions in which certain methodological rules may be suspended or replaced. The suspension of Popperian falsifiability may occur when either suitable evidential relations are absent or where there is insufficient conceptualization for deduction of possible empirical tests. This suspension allows 'breathing space' for research programs seeking evidential relations or suitable mathematical formulations, that is, it permits us to entertain as yet unfalsifiable propositions. The suspension of a rule against *ad hocness* would allow the introduction of a novel hypothesis or principle for the re-interpretation of an observation that is *prima facie* contrary to some favored belief or belief under investigation. Accordingly Copernicus introduces a second natural motion, uniform motion around the Earth's center in addition to a body's tendency to fall, to save the hypothesis of the Earth's rotation from refutation by the tower experiment. This principle appeared *ad hoc* in that initially it had no utility beyond saving Copernicus' theory from empirical falsification. However the principle could be shown to be non-*ad hoc* by the discovery of new successful applications, an achievement made later by Galileo in developing his mechanics of projectiles, pendulums, etc. Hence suspending the non-*ad hocness* rule protected the credibility of a principle that was later discovered to be highly fruitful in the production of an important family of theoretical models.

The proper conclusion to draw from Feyerabend's arguments, then, is that many, if not all, traditional methodological 'rules' are not like the constitutive rules that define a game. If one decides to suspend the rule governing the moves of a knight in chess one is suspending her effort to play that game. Rules demanding non-*ad hocness*, compatibility with observation or background belief, and empirical falsifiability are not constitutive of a 'game' of science because they can be suspended without suspending the practice of science. Rather such 'rules' are better compared to the strategies a chess player uses, such as strategies prescribing the best opening moves or for protecting the king. The game can be played without observing these strategies, and generally the strategies are effective only under certain circumstances. One who is committed to the view that science must be a rule-governed activity that can be distinguished by constitutive rules might try to defend her position by formulating second-order rules that prescribe the application of first order rules. However, without offering any argument Feyerabend dismisses this move. In the case of Copernicus' counterinductive dismissal of the tower experiment, Aristotle's principle of violent motion was only rendered suspect by the poor explanation it provided for projectile motion, and hence there was a time at which its credibility could not be appraised decisively. Hence the second order rule prescribing the application of an inductive rule is not itself decisively applicable in this case because it is not clear in these circumstances whether the Aristotelian principle is true or false. Similarly before one has discovered the fruitfulness of a principle that has been used in an *ad hoc* manner one has no way of knowing whether or not it could be made fruitful. Hence one is not always in possession of knowledge that would render the application of rules proscribing *ad hocness* decisive, and a second order rule would be useless.

In sum, it appears that methodological rules should not be compared to the constitutive rules of a game, but rather should be compared to the strategies for appraising moves allowed within the game. In general, comparing science to a game is misleading because efforts to demarcate science have thus far failed, and science apparently has no constitutive rules that govern all properly scientific endeavors in all circumstances. Even rules of deductive logic have alternatives or are subject to suspension. Nor would it improve matters by comparing science to a family of games, say one for each scientific discipline or field or for each historical stage in the development of a scientific field. Feyerabend's argument focuses on one field, astronomy, and shows that that field cannot be defined by constitutive rules. Still methodological rules may possess heuristic merit in the sense that they can, under appropriate circumstances, constrain the choices available to an inquirer so that his inquiry is more likely to be successful than otherwise. If this is the case, then it is better to understand metascience, or the 'logic' of science, in terms of the characteristics of heuristics rather than in terms of logical rules as traditionally understood, viz. as universal unquestionable and inviolable constraints on scientific activity. In this chapter we shall examine and comment on some of the methods of scientific inquiry brought up in the history of philosophy as possibly useful scientific heuristics instead of constitutive procedural rules for science.

6.2. ANALYSIS AND SYNTHESIS

The distinction between 'analysis' and 'synthesis' is drawn in many ways: The 'method of analysis' described by the Aristotelian commentator Pappus is a procedure that aids in the discovery of a proof by reasoning from its conclusion back to premises and principles that would constitute a proof.

> *Analysis*, then, takes that which is sought as if it were admitted and passes from it through its successive consequences to something which is admitted as the result of synthesis: for in analysis we admit that which is sought as if it were already done and we inquire what it is from which this results, and again what is the antecedent cause of the latter, and so on, until by no retracing our steps we come upon something already known or belonging to the class of first principles, and such a method we call analysis as being solution backwards.
>
> But in *synthesis*, reversing the process, we take as already done what was last arrived at in the analysis and, by arranging in their natural order as consequences what be-

> fore were antecedents, and successively connecting them one with another, we arrive finally at the construction of what was sought; and this we call synthesis.
>
> Now analysis is of two kinds, the one directed to searching for the truth and called *theoretical*, the other to finding what we are told to find and called *problematical*. (1) In the *theoretical* kind we assume what is sought as if it were existent and true, after which we pass through its successive consequences, as if they too were true and established by virtue of our hypothesis, to something admitted: then *(a)* if that something admitted is true, that which is sought will also be true and the proof will correspond in the reverse order to the analysis, but *(b)* if we come upon something admittedly false, that which is sought will also be false. (2) In the *problematical* kind we assume that which is propounded as if it were known, after which we pass through its successive consequences, taking them as true, up to something admitted: if then *(a)* what is admitted is possible and obtainable, that is, what mathematicians call *given*, what was originally proposed will also be possible, and the proof will again correspond in the reverse order to the analysis, but if *(b)* we come upon something admittedly impossible, the problem will also be impossible. (Pappus of Alexandria, *Mathematical Collection* VII, 1–3, Quoted in Cohen and Drabkin, 1952, p. 39)

Pappus' use of "consequences" suggest that the steps of the reasoning he is considering are reversible so that premises are as much consequences of the conclusion to be proved as conversely.

In mathematics such reversibility can occur if the reasoning is in terms of equal quantities, such as algebraically equivalent representations of sums and products, equal lengths, areas and volumes in geometry, or if it is in terms of geometrical similarities. The effective application of a similar procedure in a system of natural deduction for, e.g., sentential calculus, indicates that this condition of reversibility is not necessary for the method to have heuristic merit. In this application the procedure is to look for the kind of principle that could be used in the proof of the conclusion by considering logical connectives in the conclusion and items already admitted as premises or as theorems. Once a plausible inference principle is found one can appeal to its form and the form and content of the conclusion to determine what premises are needed in its application for the conclusion's proof. If these premises are yet to be established, their proof can be sought by the same procedure until established propositions or principles are reached. In effect, the 'theoretical kind' of analysis described by Pappus leads to a question, viz. whether or not the proposition generated by the analysis is true.

In sum, the heuristic efficacy of this method depends first upon using the form and content of the conclusion to suggest possible evidential linkage to what has already been established. Second, Pappus' description of the procedure as leading to possibly false premises suggests that analysis can also be used to discover possible evidence whose acceptability has yet to be determined. Hence as a search for evidence, the method of analysis need not be considered biased toward the truth of the conclusion. Both of these features appear in the investigative heuristic of entertaining possible answers to a problem in search of evidence to be sought or determined that was described in earlier chapters. This similarity to an effective deductive heuristic in respect to what gives the latter heuristic its strength, particularly its power to bring to the investigators attention principles that can be relevant to the proof, can be used in arguing for the power of the investigative heuristic.

It has been suggested that in addition to being regression to axioms and postulates needed for a proof, traditional geometrical analysis involves breaking geometric figures down into components (Hintikka and Remes 1974, Hintikka, 1978) such as circles, semicircles, chords, tangents, triangles, proportional and parallel lines. Correspondingly, 'synthesis' is the procedure of assembling relevant principles and premises, and possibly components and relations among the components, all of whose discoveries are facilitated by analysis, to constitute the finished proof. In the seventeenth century Newton gave this distinction what would appear to be a broader interpretation:

> As in mathematics, so in natural philosophy, the investigation of difficult things by the method of analysis ought ever to precede the method of composition. This analysis consists in making experiments and observations, and in drawing general conclusions from them by induction, and admitting of no objections against the conclusions but such as are taken from experiment, or other certain truths. For hypotheses are not to be regarded in experimental philosophy. And although the arguing from experiments and observations by induction be no demonstration of general conclusions, yet it is the best way of arguing which the nature of things admits of, and may be looked upon as much the stronger by how much the induction is more general. And if no exception occur from phenomena, the conclusion may be pronounced generally. But if at any time afterward any exception shall occur from experiments, it may then begin to be pronounced with such exceptions as occur. By this way of analysis we may proceed from compounds to ingredients and from motions to the forces producing them, and in general from effects to their causes and from particular causes to more general ones, till the argument end in the most general. This is the

> method of analysis; and the synthesis consists in assuming the causes discovered and established as principles, and by them explaining the phenomena proceeding from them and proving the explanations.
>
> In the first two books of these *Optics*, I proceeded from this analysis to discover and prove the original differences of the rays of light in respect of refrangibility, reflexibility, and color, and their alternate fits of easy reflection and easy transmission, and the properties of bodies, both opaque and pellucid, on which their reflections and colors depend. And these discoveries being proved may be assumed in the method of composition for explaining the phenomena arising from them, an instance of which method I gave in the end of the First Book. In this Third Book I have only begun the analysis of what remains to be discovered about light and its effects upon the frame of nature, hinting several things about it and leaving the hints to be examined and improved by the further experiments and observations of such as are inquisitive. (Newton, 1730, pp. 404f)

Among the several things Newton attributes to the 'method of analysis' is experimenting and observing, drawing general conclusions by induction, 'proceeding' from compounds to ingredients, motions to forces, effects to causes, particular to more general causes. "Proceeding" used here seems multiply ambiguous: It could be a kind of inference, e.g. from properties of compounds to those of their ingredients, from motions or other effects to their causes, from causes acting in a narrower to causes acting in a broader domain. Secondly, Newton's words indicate that "analysis" could designate the experimental actions of separating ingredients, as Newton experimentally separated the ingredients of white light for further experimentation and observation. Similarly one might seek to separate for observation the components or 'principles' supposed to be in chemical compounds. It might also be a kind of intellectual division of some process into components, as Newton began, in his mechanical treatment of planetary motion, with a theoretical model of a planetary system stripped of several components and features the Solar system is known to possess. His initial model consists of a point-like planet moving about an immovable force center. (Cf Cohen, 'the Newtonian Style', 1980, Chapter 3.) Furthermore 'analysis' might be construed as experimental efforts to separate, similarly to the analysis leading to the initial planetary model, the essential components of a causal process from extraneous disturbances by a contingent environment and from random or uncontrolled fluctuations. Newton's 'analysis' can also be construed as an empiricist's version of Descartes' analytic method of regression to presuppositions. Descartes' method searches for self evident and indubitable first principles, such as those involving thought and the existence of the self as thinking substance, the perfection of God as presupposed by the idea of a perfect being and the extension and motion of matter as presupposed by material existence and differentiation of individual material things. Descartes' analysis is analogous to regression to self-evident first principles in geometry and to the method offered by Pappus. However, Newton demands that the first principles of natural philosophy, or *vera causae*, be established by experiment and then extended in their application by induction. It would appear that this inductive extension is the only plausible means for inferring causes, such as centripetal forces, from effects such as the motions of planets in closed planar orbits. If Newton's second law of motion is synthetic, there is no deduction from accelerations as effects to forces as causes unless the truth of the law is supposed as a major premise.

Finally, Newton's "synthesis" can be interpreted by looking at what appears at the end of his Book I of the *Opticks*: There the properties of 'simple homogeneous light' are appealed to in explaining the appearances produced by the prism, the rainbow, and the permanent colors of various objects. Another example occurs in Book III of the *Principia*, where he applies much that was discovered in Book I in regard to orbits in centripetal forces, orbits in inverse square central forces, two-body problems, bodies occupying extended volumes, etc. to the 'system of the world', i.e. the Sun, planets and their satellites as generally known in the then current cosmological background. In sum, Newton's method of 'synthesis' is less ambiguous than his 'analysis' and involves the construction of explanatory models. These models are based in part on the laws and principles produced by the initial application of the method of analysis and in part upon further information about real world systems, such as rainbows and planetary systems.

Still, Newton's own version of the method of analysis suggests several different kinds of heuristic that seem, Hintikka *et. al.* to the contrary, to be quite distinct: (i) regression in search of evidential relations and evidence, (ii) reductionism in which knowledge of component is a means to knowledge of compound, (iii) abstraction of essential mathematical geometric or mechanical characteristics from others.

Hintikka *et. al.* go to great lengths to point out that Pappus' examples of analysis in geometry are not simply regression to premises or presuppositions, but also include reducing a given figure to its components, where the components and their mutual relations as well as auxiliary constructions contribute heuristically to the discovery of the proof. This, they claim, is distinct from the method of analysis found in Aristotelian logic which involves breaking down arguments into component but known syllogisms. This last procedure includes no actual separation of components from composites. In the example they consider (1974, pp. 23ff) a circle ABG with center E is given with a diameter BG extended to join a tangent at point D, and an arbitrary line from D is extended to join the near circumference of the circle at J and the far circumference of the circle at F (Figure 2). Diameter AH is drawn to tangent point A and H is joined to F with a straight line intersecting the first diameter at K. Also H is joined to J by a straight line intersecting the first diameter at L. The conclusion to be proved is that segments on the first diameter KE=KL, and this conclusion is assumed true in the first step of the analysis. In the proof two

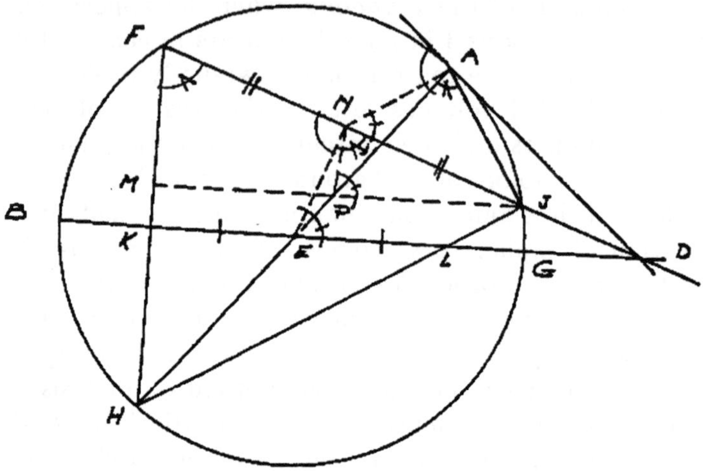

Figure 2: From Hintikka and Remes (1974) p. 24.
Reproduced here by kind permission of Kluwer Academic Publishers.

auxiliary constructions are added, JM drawn parallel to the first diameter BG and a perpendicular from the center of the circle E to the line DF. The components of this figure include the given center E, the diameters BG, AH, the tangent AD, chords FJ, AJ. Further components emerge once these lines are drawn and auxiliary constructions added, e.g. triangles HKE, HEL, HMP, HPJ with respectively parallel bases KE, MP and EL, PJ. It is by appeal to the properties of triangles with parallel bases that the first equality MP=PJ is inferred from the assumed conclusion KE=EL. That is, component figures, whether or not formed in combination with auxiliary constructions, have properties that can be known from previous postulates or theorems, and these properties may be useful in the search for proofs, whether the search be regressive from the conclusion to be proved or progressive from known theorems to the conclusion. Similarly properties of chords of circles show that the chord FJ is bisected by the perpendicular to the diameter EN, so that FN=FJ. Triangles FJM, NJP are formed by the given lines and auxiliary constructions and from known properties of triangles whose sides FJ, MJ are bisected by NP, it follows that bases NP and FM are parallel. From this follows the equality of angles JNP, NFM. Like properties give the equality of angles JNP, NFM. Also properties of tangents imply that angle EAD is a right angle.

In this case the heuristic advantage of the spatial analysis is that component tangents, parallels, triangles, etc. have known geometric properties. These properties, possibly in the light of their mutual relations within the given figure, may provide inferences from the conclusion to be proved to known postulates or theorems, or may themselves be known as theorems or postulates. The theorems and postulates thus revealed will serve as premises in the synthesis and either the inferential steps can be reversed or can suggest converse inferences to be sought in completing the synthesis.

Correspondingly, Newton's experimental analysis of white light into component colored light with a prism enabled him to attribute values for the 'refrangibility' of each color in, e.g., water by passing through water the various colors thus produced. This procedure can be considered one of inferring from effect, the separation of colors by a prism, to a cause, the differing refrangibilities of various forms of light. However, the inference is not really direct, for Newton has to exclude other possible causes, e.g. the possibility that the prism modifies the light rays. He excludes this latter possibility by another experiment, that demonstrating that once colored light is separated its color and refrangibility are not altered by further passages through prisms. This procedure, like the induction described above, is a subheuristic to Newton's analysis that is not found in geometry but is, as examples discussed earlier indicate, effective and hence valuable in investigations in natural science and other enterprises in which the discovery of causes is important. This particular analytic procedure also is a heuristic for separating essential causal relations from peripheral contingencies. The composition of white light from one or another source is just such a contingency and exact laws of refraction cannot be attributed to these instances without Newton's analysis into component colored light. Similarly, exact laws of motion cannot be formulated without analyses that reduce or eliminate the peripheral contingencies of air resistance, sliding, rolling and bending friction. However because an instance of analysis has both the heuristic power of reducing an unknown whole to known or knowable components and the power of separating accidental from essential causes does not mean that all instances of analysis will have both of these heuristic characteristics. Other instances can still reduce unknown wholes to knowable parts without dealing with peripheral contingencies, as the above geometrical example illustrates. Similarly the inductive extension and the exclusion of alternative possible causes are subheuristics that may or may not be joined to an analytic procedure. Furthermore, Newton's procedure is not exactly analogous to Pappus' example of geometrical analysis. The rainbow is presumed to occur in a configuration of white light source and refracting bodies that is quite different from the configuration of water prisms in the initial analysis. In this latter configuration several raindrops are considered as prisms and internal reflectors lying between the white light source, the Sun, and the observer. Values for the refrangibility of various colors are applied to this geometric configuration in the effort to construct lines of sight from which the observer will see different colors, and these lines of sight will give the configuration of the apparent rainbow. Of course Pappus or other geometers can apply properties of tangents, parallels, chords, triangles, etc. in the construction of other proofs, but in natural science the epistemic value of synthesis arises out of the application of principles, laws, or properties of components to configurations other than those from which these principles, etc. were discovered. Using the refrangibility of light discovered by two refractions in a given pair of water filled prisms to explain refraction in these very prisms is a trivial endeavor in natural science. It is by extending properties thus discovered to different configurations that empirical or explanatory consilience is achieved.

The deductive structure of geometry and physics are comparable in this sense: The postulates of Euclidean geometry are about points, lines and planes, and contain no description of figures such as that mentioned in the above example drawn from Pappus' writings. Rather the figure considered in the example is given in the Proposition. These givens and their properties can serve, in addition to axioms, postulates and previously proven theorems, as premises in the proof or construction sought. Similarly, in mechanics the laws of motion, gravity, etc. do not describe the configurations of planets in the solar system, the initial and boundary conditions pertinent to the motion or a projectile, or the like. These systems, initial and boundary conditions are givens which partially distinguish each problem. These given configurations, according to Hintikka, *et. al.* are heuristically important in the analyses occurring both in geometry and in mechanics, a point that is readily demonstrated in the examples considered above. However, they also claim that analyses in syllogistic do not appeal to such configurations and the implication of this point may be that the deductive structure of logic is quite different from that of geometry and mechanics. In logic theorems are proved directly from axioms without the givens or systems with initial and boundary conditions. Hence the whole heuristic of breaking unknown configurations into known components from which something new can be discovered in mutual relations of components has no application in logic. However, still in natural deduction there is a kind of configuration: A proof is a one-dimensional sequence of steps that are so related that the premises of the subsequent are conclusions of prior inferences. Within these sequences are components that are direct applications of axioms, such as an occurrence of *modus ponens* or a distribution of conjunction through disjunction. However it is also important for the student of natural deduction to build up a repertoire of routines that are not direct applications of what are taken as axioms, such as reducing conditionals to disjunctions for the purpose of applying commutative, distributive and DeMorgan rules for producing desired rearrangements of formulae. These strategies are appropriate only when placed in a context of premises and conclusions where they can provide intermediate deductive links needed in completing a proof. Hence logical deductions have components with known or knowable properties, and these components, as in mechanics and geometry, are heuristically useful only when they have, or there can be discovered appropriate relations to premises and conclusions. Furthermore natural deduction is at least as frequently applied to contingent 'givens', premises that are not logical axioms or theorems, as it is applied in proving logical theorems. Hence it cannot be denied that logic, like mechanics and geometry, also deals with items that do not appear in the axioms. Again, the structure of these 'givens', e.g. whether they are conditionals, conjunctions, disjunctions, negations, universal or existential, is an essential component of all of the effective heuristics used in natural deduction. Thus the heuristic reduction of unknown whole to known components and configurations is as important to discovering proofs in logic as it is in geometry.

In conclusion, the method of analysis described by Pappus, Descartes, Newton and others is not a single heuristic. Rather it is a family of heuristics and sub-heuristics which can most generally be described as having three distinct components: (i) By appealing to characteristics of a conclusion one may be in a position to infer the nature of the evidence required for the conclusion. Sometimes circumstances are such that these inferences are deductive and reversible, but in other cases they may only provide a range of possible steps that is more limited than would occur without the regression. Also conclusions may only provide the weakest hints as to possible evidential relations leaving the investigator with the task of establishing what could count as evidence as well as the evidence itself. Such circumstances are common in natural science where it is not initially known what kind of evidence would bear upon the truth or falsity of a hypothesis. The search for evidential relations is as much a part of the discovery process in natural science as is the search for suitable ideas or concepts. (ii) Material objects, geometrical figures and even logical arguments have components, and in some epistemic circumstances these components and their relations may be well known. However, there may also be various sub-heuristics for discovering these components. In geometry they can often be discovered by inspection, though the addition of auxiliary constructions, as Hintikka *et. al.* point out, may help in finding heuristically important relations among the components. In logic inspection of the structure of premises and conclusions and combining regression from the conclusion with progression from premises can reveal the known components, whether axiomatic or derivative, that when appropriately assembled will constitute the proof. The subheuristics for finding the components may differ in logic from those in geometry. Still the reductive heuristic plays analogous roles in the two disciplines, a strategy of knowing the whole by means of knowledge of parts and their mutual relations. Similarly, in mechanics components of a system are not always discoverable upon inspection. A variety of sub-heuristics are needed in this discipline that would not be appropriate in geometry or logic. Some of these are the experimental heuristics mentioned by Newton for separating white light into its components or chemical compounds into components. In this case the components may not be known initially, but they may also be knowable only as components, as would be the case if their behavior could not be isolated from peripheral contingencies without actual physical separation. Accordingly, physical or experimental isolation is a subheuristic for reductionistic analyses particularly appropriate in the natural sciences. Secondly the theoretical heuristics of considering artificially simplified models initially to establish laws or principles, and then to approach models of real world systems by successive introduction of more realistic circumstances in subsequent models was used by Newton with profound consequences and combines heuristics broadly described above as analysis and synthesis. Also, as will become more apparent in the section to follow, various hypothetical and inductive heuristics can be used as a means of arriving at knowledge of components. These can be an alternative sub-heuristics to theoretical and experimental analysis in reductionistic strategies. Finally, reductionistic strategies may fail where essential causal relations are multiple and complex or not clearly distinguishable from peripheral contingencies (Wimsatt, 1974, 1976, 1979, 1980). (iii) The method of isolation for separating peripheral contingencies for essential causes has also been called analysis, though we have seen it can be a component of something else, reduction, which is also called analysis. This method of isolation can be regarded as a heuristic for discovering clear causal laws that has been successfully applied in mechanics and optics, as examples mentioned above indicate. It also has an analog in heuristics for constructing mathematical models, such as Newton's models of planetary systems. It also need not be a sub-heuristic of some reductionistic heuristic, for Galileo's use of it in analyzing motions of projectiles, pendulums, etc. is not reductionistic in the sense just discussed.

6.3. 'INDUCTIVE' HEURISTICS

As in the case of "analysis", "induction" has a variety of usages in the philosophical tradition. For Aristotle it includes a kind of intuitive *perception* of the universal in the particular, and is not just an inference based on enumeration (*Posterior Analytics*, II, 19). For others the term designates the elementary enumerative induction described by Hume as an item of human and animal psychology upon which humans could offer little improvement. Mere repetition of observed instances of, e.g. A accompanying B, enhances the credibility of unobserved instances of the same occurrences or of the universal 'All A are B'. Though Hume adds additional requirements of spatial contiguity and temporal succession to inductions to causal relations, their credibility remains directly proportional to the number of repetitions observed. However, the several methods ascribed to Francis Bacon, John Herschel and John Stuart Mill are also described as 'inductive' but do not appeal just to repetition of instances in ascribing credibility: In the 'method of agreement' credibility depends upon the elimination of possible causes C,... by demonstrating instances in which C,... are absent but A and B remain present. Here it is assumed that causes must be at least necessary conditions for their effects. However trial removals of accompanying conditions C... which are distinct from A can give grounds for believing that A is also sufficient for B, and in this case the credibility of the conclusion depends on the number of accompanying conditions C,... examined. Hence enumeration can be part, but not the only part of the epistemic basis of Mill's canon of agreement. The 'method of difference' is said to ascribe credibility to those factors that cannot be eliminated by agreement. If one tries to eliminate A but in the process also eliminates B, the credibility of A's causing B is enhanced. Again, the method of difference shows that A is at least part of the cause of B, that is the presence of A is necessary for the occurrence of B, though there may also be other necessary conditions for B that are not uncovered. The 'method of concomitant variation' bases credibility upon covariation of quantities in A and B, again not necessarily upon the number of instances examined. Finally the 'method of residues' bases credibility upon causal relations already known in a given circumstance to be exhaustive of the causal circumstances save one condition and effect. Again the strategy here is elimination, not enumeration, and what is eliminated are known causal relations leaving the co-occurrence of A and B as a residue. One further family of methods commonly described as 'inductive' involve the fitting of curves to data 'points'. In this case credibility is based upon the density of the data points, the precision of the measurements upon which the points (actually rectangles) are based, and the algebraic 'simplicity' of the function whose plot connects them. Increasing precision and density of data again eliminates curves not consistent with the data, and once again the strategy for enhancing credibility is the elimination of alternatives.

In discussing Kepler's discovery Mill denies that Kepler's procedure, which Mill describes as one of fitting observed positions to a curve, is inductive. Mill wants to restrict the term to processes involving generalization, but he believed that generalizations about past and future planetary orbits preceded Kepler's contributions (1893, Book III, Chapter II). Inductions for Mill were also conceptually barren. They do not generate or depend upon the application new conceptions or new viewpoints, that is, conceptions beyond those already found or applied in the observations. By contrast, Whewell (1849) argues that Mill's restriction of induction to bringing together facts in temporal relation is arbitrary and should not exclude Kepler's bringing together facts in spatial relation or Newton's bringing them together in relation to force. Also he denies the conceptual barrenness of induction. New conceptions, or conceptions beyond those needed for initially given facts, are needed for their 'colligation', that is, their relation under a new viewpoint. The novelty of such conceptions is not total, as would be the revolutionary conceptual novelties of Feyerabend and Kuhn. For Whewell novel concepts are developments upon prior Kantian archetypal ideas. Thus Kepler's colligation of observed locations of Mars under the new conception 'ellipse', or at least a conception newly applied in these circumstances, is a proper induction. In this case credibility is based on 'colligation' of facts under a new organizing perspective and 'consilience' of empirical evidence. 'Consilience', as described earlier, occurs when initial items of evidence are confirmed by a number of further items of evidence from independent sources that support the same conclusion.

'Inductivism' is plausibly construed as the view that the 'inductive' methods of Hume and Mill, including methods of curve fitting and rounding off experimental errors, but exclusive of those of Whewell, are exhaustive of procedures for generating scientific knowledge. Inductivism also includes several assumptions about 'facts' and 'observations', particularly the empiricist assumption that science is based upon observable facts that can and should be discovered and communicated independently of any theory. Secondly, inductive empiricists assume that once properly established such observed facts enjoy maximum credibility and no fallibility. Such facts serve as the ever accumulating foundation for all science.

Thus formulated, inductivism restricts scientific knowledge to phenomenalistic correlations and generalizations. Accordingly proper scientific laws and theories can be expressed only in terms accessible to observation, or in terms suitable for the description of the foundational observable facts. Astronomy is restricted to observing, computing and plotting various periodicities in planetary movements, including visibilities, longitudes, latitudes, conjunctions, oppositions and retrogressions, and predicting such events on the basis of these computations. Hypothetical mechanisms for 'explaining' these motions, such as celestial spheres, motive forces from the Sun or universal centripetal gravitational forces are excluded from scientific descriptions and explanations. Genetics is restricted to correlations between the history of a line of descent, including whether it remains in-bred or crossed with other lines, and the ratios in which various possible traits might occur in these lines. For example, blue eyes might occur consistently in a line of humans never crossed with those with brown eyes, but when such crossings occur all brown eyes might appear in the first generation and blue and brown in a ratio of 1:4 might appear when members of this first generation interbreed. Excluded are 'factors' or 'genes' that occur in pairs in each organism which can be described as causes of whatever traits are expressed but may occur without observable expression if they are described as 'recessive'. Inductivism admits 'laws' that are expressed as mathematical functions which also fit a distribution of rectangles (data 'points') in which experimentally inexact 'data' are plotted. Thus the relation of velocity to elapsed time for bodies observed rolling down an inclined plane might be represented as a linear function whereas the relation of distance to elapsed time would be represented as a quadratic function. However an account in terms of a 'mechanism' of proximate or generative causes, that is entities, states or events whose fundamental 'nature' it is to act in certain ways under certain circumstances, is excluded. Inductivism excludes a theory of gravity that describes this force as essentially and fundamentally a form of action at a distance between massive bodies whereby mutual centripetal accelerations are produced. Similarly it would exclude a field theoretic account where gravitation is a essentially and fundamentally a manifestation of a condition, the intrinsic curvature of a four dimensional space-time metric, that is continuously distributed throughout the Universe.

That inductivism excludes causal mechanisms and other explanatory hypotheses that are considered essential to past and present scientific theories is a powerful and persuasive refutation of this form of inductivism. However, the correlational and other 'inductive' methods just described remain an important component of the practice of science. Hempel's (1966) description of Ignaz Semmelweis' isolation of contact with 'putrid matter' between 1844 and 1848 as a cause of elevated rates of childbed fever in certain maternity wards at the Vienna General Hospital aptly illustrates the utility of Mill's methods of agreement and difference in certain epistemic circumstances. Epidemiological research of this kind occurs when one searches for causal factors in the absence of knowledge of suitable mechanisms for the transmission and etiology of diseases. Semmelweis held or entertained initial beliefs about certain causal factors, e.g. that they are caused by 'epidemic influences' or 'atmospheric-cosmic-teluric changes' that spread over whole geographic regions, or by overcrowding in the wards. These causes are eliminated because they fail to explain the localization of the epidemic in the first ward and the difference in the occurrence rates of the disease in the two equally crowded maternity wards under his observation. A second circumstance appropriate for the application of these methods is that there be a number of plausible causal hypotheses that can be tested and excluded by Mill's methods. Thus Semmelweis excluded epidemic influences, overcrowding, rough treatment by interns, position in giving birth, and fear of death by a combination of the methods of agreement and difference. The condition that could not be excluded, contact with cadavers before handling maternity patients, was suggested to Semmelweis by a physician's accidentally contracting the disease from a minor cut from a dissecting scalpel. (Hempel, 1966, pp. 4f) Thus a correlate with the increased rate of the disease was isolated and this isolation was sufficient for the limited practical propose of bringing down the elevated rate in the second ward. This practical objective is clearly a good reason for Semmelweis' highlighting the problem raised by the difference between the two wards. However, his correlational methods it did not produce knowledge of a mechanism, such as bacterial infection. It appears that this case demonstrates that these correlational methods are heuristics and not constitutive rules for the conduct of science, as would be the claim of the version of inductivism described above. These methods are appropriately applied in circumstances in which correlations and not fundamental mechanisms are sought, though their application might well be effective in preparing investigators for seeking such mechanisms. They cannot be applied in an epistemic vacuum, that is in contexts of causal agnosticism, as inductivists often demand, but are effective only in the exclusion of plausible causes. The more the possible causes are narrowed by approximately correct plausibility judgments the more heuristically effective the methods are, for the task of excluding conjectures with no prior epistemic merit can be indefinitely large. Correlational claims have practical value in medicine in that they can be used to control epidemics, but if medical science viewed correlational methods as constitutive rules it would lose the practical power of preventing diseases by immunization or intervening with the progress of an infection with antibiotics.

In his methodological remarks Newton also used the term 'induction', as should be apparent from the passage quoted in the preceding section. However, Newton's 'Rules of Reasoning in Philosophy' shed further light upon what might be considered to be another kind of 'inductivism':

> **RULE I**
> *We are to admit no more causes of natural things than such as are both true and sufficient to explain their appearances.*
>
> To this purpose the philosophers say that Nature does nothing in vain, and more is in vain when less will serve; for nature is pleased with simplicity and affects not the pomp of superfluous causes.
>
> **RULE II**
> *Therefore to the same natural effects we must, as far as possible, assign the same causes.*
>
> As to respiration in a man and in a beast, the descent of stones in Europe and America, the light of our culinary fire and of the sun, the reflection of light in the earth and in the planets.
>
> **RULE III**
> *The qualities of bodies, which admit neither intensification nor remission of degrees, and which are found to belong to all bodies within the reach of our experiments, are to be esteemed the universal qualities of all bodies*

whatsoever.

For since the qualities of bodies are only known to us by experiments, we are to hold for universal all such as universally agree with experiments, and such as are not liable to diminution can never be quite taken away. We are certainly not to relinquish the evidence of experiments for the sake of dreams and vain fictions of our own devising; nor are we to recede from the analogy of Nature, which is wont to be simple and always consonant to itself. We in no other way know the extension of bodies than by our senses, nor do these reach it in all bodies; but because we perceive extension in all that are sensible, therefore we ascribe it universally to all others also. That abundance of bodies are hard we learn by experience; and because the hardness of the whole arises from the hardness of the parts, we therefore justly infer the hardness of the undivided particles, not only of the bodies we feel, but of all others. That all bodies are impenetrable, we gather not from reason, but from sensation. The bodies which we handle we find impenetrable, and hence conclude impenetrability to be a universal property of all bodies whatsoever. That all bodies are movable and endowed with certain powers (which we call inertia) of preserving in their motion, or in their rest, we only infer from the like properties observed in the bodies which we have seen. The extension, hardness, impenetrability, mobility and inertia of the whole result from the extension, hardness, impenetrability, mobility, and inertia of the parts; and hence we conclude the least particles of all bodies to be also all extended, and hard

> and impenetrable, and movable, and endowed with their proper inertia. And this is the foundation of all philosophy. ...
>
> Lastly, if it universally appears, by experiments and astronomical observations, that all bodies about the earth gravitate toward the earth, and that in proportion to the quantity of matter which they severally contain; that the moon likewise, according to the quantity of its matter, gravitates toward the earth; that on the other hand, our sea gravitates toward the moon; and all the planets one toward another; and the comets in like manner toward the sun: we must, in consequence of this rule, universally allow that all bodies whatsoever are endowed with a principle of mutual gravitation. For the argument from the appearances concludes with more force for the universal gravitation of all bodies than for their impenetrability, of which, among those in the celestial regions, we have no experiments nor any manner of observation. ...
>
> **RULE IV**
>
> *In experimental philosophy we are to look upon propositions inferred by general induction from phenomena as accurately or very nearly true, notwithstanding any contrary hypothesis that may be imagined, till such time as other phenomena occur by which they may either be made more accurate or liable to exceptions.*
>
> This rule we must follow, that the argument of induction may not be evaded by hypotheses. (From the *Principia*, Vol. III, quoted in Thayer and Randall, 1953, pp. 3ff)

First, these rules concern 'inductions' in the sense that they are designed to warrant the extension of what can be established in experimentally and observationally accessible domains, what falls "...within the reach of our experiments", to other domains. These other domains, for which "...we have no experiments nor other manner of observation", include the "celestial regions" as well as small bodies that we can actually divide and those that can be distinguished only by our minds. Rules I and II include an effort to 'justify' these extensions: According to Rule I, were Nature to produce a given effect in one domain by cause C and in another by a different cause C′, it would be employing more which "...is in vain when less will serve...". Rule II expresses a kind of uniformity of nature, viz. a uniformity of causal relations, and Newton indicates in its statement that he believes that it is a consequence of the kind of simplicity described in Rule I. Rule IV states that this simplicity and uniformity is only a defeasible presumption, not a metaphysical certainty.

Newtonian induction does not restrict scientific inquiry to phenomenalistic generalizations and correlations as does the variant of inductivism described earlier. Newton claims to have 'deduced' his first principles or axioms, the 'Laws of Motion' from 'phenomena' (letter to Cotes, quoted in Thayer and Randall, 1953, p. 6). The axioms are expressed as mathematical functions that have been interpreted as representing fundamental causal processes, though Newton himself was never satisfied that he hand fully understood the mechanism of gravity (Hesse, 1962). The relevant phenomena are described in the *Principia* as observations from experiments performed by Galileo and Huygens.

> Hitherto I have laid down such principles as have been received by mathematicians, and are confirmed by abundance of experiments. By the first two laws and the first two Corollaries, *Galileo* discovered that the descent of bodies varied as the square of the time ... and that the motion of projectiles was in the curve of a parabola; experience agreeing with both, unless so far as these motions are little retarded by the resistance of the air. When a

> body is falling, the uniform force of its gravity acting equally, impresses, in equal intervals of time, equal forces upon that body, and therefore generates equal velocities; and in the whole time impresses a whole force, and generates a whole velocity proportional to the time. And the spaces described in proportional times are as the product of the velocities and the times; that is, as the squares of the times. (1962, Vol. I, Scholium to Law III. Also Vol. II, p. 408, contains references to Huygens' experiments with pendulums.)

As Newton acknowledges, Galileo showed that the gravitational components, i.e. the vertical components, of projectile motions are such that the distance traveled is as the square of the elapsed time. Newton notes that this principle is a consequence of the assumptions that uniformly accelerated velocity is proportional to elapsed time, i.e. $v_f = gt$, where 'g' denotes the uniform acceleration of gravity at the earth's surface. If in small time intervals the applied force acts in impulses, these impulses generate in those times small increments of velocity. If the impulses are always the same, by Law II the increments of velocity are also the same and the number of increments accumulated over a period of time, the sum of these small time intervals, will be proportional to that period of time. He also adds that the distance covered by a moving object is the product of its average velocity and its time, i.e. $s = v_{avg}t$. Combining these propositions in view of the relation $v_{avg} = 1/2(v_o + v_f)$ gives Galileo's squared time rule as a consequence when the initial velocity $v_o = 0$, as in free fall from a fixed position. Thus $s = v_{avg}t = 1/2(0 + v_f)t = 1/2gt^2$. However, if the distance is observed to be as the square of the elapsed time $s = Kt^2$, as Galileo observed in various experiments with inclined planes, it can be deduced, conversely to the above deductions, that uniform acceleration is taking place. That is, for a given plane K is observed to be constant and $s = Kt^2 = Kt(t)$ in repeated experiments. This equation agrees with the above $s = v_{avg}t = 1/2(v_o + v_f)t$, which is Galileo's law of mean velocities combined with the kinematic definition of uniform velocity, provided $Kt=1/2(v_o+v_f)$. As a condition of the experiment $v_o=0$, and $v_f=at$ by Newton's analysis and integration cited above. Hence $Kt=1/2at$ or $K=1/2a$, and hence the acceleration a for this experiment must be constant if the observed K is constant.

One might object that Galileo's experiments provide only fuzzy data, that is data within intervals of uncertainty for both the elapsed times and the distances that are co-measured. Accordingly the data do not form a series of points sufficiently dense to determine uniquely the function $s=Kt^2$. Rather they form a series of rectangles representing intervals of uncertainty that may be linked by an infinite variety of variously curved lines representing a variety of different functional relations between distance and time. Hence there can be no deduction from actual measurements, which can never quite eliminate the intervals of uncertainty in both spatial and temporal variables. There may be deductive methods of eliminating some functions that are incompatible with a set of data-rectangles, but such methods cannot be expected to be exhaustive of all possibilities so that only one believable function remains compatible with the data. Inductive methods of eliminating remaining alternatives by increasing precision still leave the functions sought underdetermined. Accepting $s=Kt^2$ on grounds of its 'simplicity' has no epistemic import because there is no prior assurance that mechanical behaviors can be represented by relatively simple functions such as linear or quadratic functions.

Still it might be claimed that once one has come to grips with doubts about inductive reasoning from available observational data with its inevitable intervals of error one can accept that $s=Kt^2$ as an empirical law. Newton can be said to accept this law as Galileo's conclusion from which, then, he can deduce that acceleration is constant or that $v_f=at$. This procedure is comparable to Newton's reasoning from Kepler's three laws, from which follows planetary motion under central force and the inverse square law. However, Kepler's laws can hardly be said to have been 'deduced from the phenomena'. The former deduction can be regarded as instantial for a given inclined plane, but then its conclusion can be extended to other inclined planes and to free fall 'by induction' in Newton's sense. To claim that different causes are operating for other inclined planes $i=2,\ldots n$ for which $s=K_it^2$ would violate Rule I, and these $n-1$ planes can include one whose inclination is vertical, which is free fall. In these cases different observed K_i would be equated with different accelerations a_i and for the vertical one $a_i = g$.

However it appears that Whewell is right in pointing out that these empirical laws of Galileo and Kepler are not strictly derived from sensory data but are constructs that make use of 'colligating ideas'. In Galileo's case the ideas are mathematical conceptions of uniform acceleration and its deductive consequences, the vertical natural acceleration we call free fall and the conception of the inclined plane as providing a constraint that reduces the acceleration of free fall by the sine of the plane's angle of inclination to the horizontal.

Feyerabend (1962) and Lakatos (1978) have argued that because Galileo's and Kepler's laws are shown false by Newton's gravitational theory Newton could not have reasoned deductively or inductively from these laws to his theory. However, both overlook Newton's heuristics, which include his application of the method of analysis to problems of planetary motion, as described briefly above. Newton's inferences regarding celestial gravity include a judicious combination of the method of analysis with the inductive extension laws from fully empirically accessible domains to domains that are partially empirically accessible. Thus one can reason that planets move under centripetal forces by observing in terrestrial experiments that bodies moving in circles, e.g. a stone in a sling, must be restrained by a centripetal force to continue so moving. In the case of the planet the near circular path of the planet is observable but the centripetal force is not, though both are observable in the terrestrial experiment. Rule II then allows the inference of centripetal forces constraining satellite motions. Newton also deduces this conclusion from Kepler's Second Law, the 'areas law', thereby providing a kind of consilience for this hypothesis. The form of the law, the inverse square, is deduced from Kepler's third law, the 'harmonic law' that $T=KD^{3/2}$, though again under the simplifying assumption that planetary and satellite orbits are circular. However, nothing as yet follows about the identity of this celestial centripetal force. Newton provides a continuity argument whereby he demonstrates that a terrestrial satellite is simply a projectile with appropriate initial positions and velocities. Hence if projectiles are constrained by terrestrial gravity, so are terrestrial satellites, including the Moon. Since the Moon, the planets and other planetary satellites all obey Kepler's Third Law, they are also constrained by what we call gravity, though in this case solar and planetary gravity. This identification of celestial centripetal forces with terrestrial gravity allows an inference in the other direction: Since it has already been deduced from Kepler's Third Law that celestial centripetal forces are inverse square, terrestrial gravity must be so also. This last conclusion was confirmed by Newton's calculation of the centripetal acceleration of the Moon from the inverse square law applied to the mean distance of the Moon from the Earth, which produced a value that agrees with the known curvature of the Moon's orbit. As in the case of the first of Newton's models of planetary motion, these inductions and deductions are performed on a simple but approximate model of the Solar System, viz. one in which the sun is stationary and in which Kepler's second and third laws are strictly true because the effects of the Sun's mobility and planetary gravitational interactions are not taken into account. A second approximation in this model neglects Kepler's first law by assuming the orbits to be circular, an assumption that simplifies the deduction of the inverse square law. This initial model is believed to be false on grounds of Newton's third law, which requires that the sun and the planets move about a common center of gravity, on grounds of his law of gravity, which requires that a given planet's motion be influenced by certain other planets as well as by the Sun, and on grounds of Kepler's first law. From these laws it is anticipated that these departures from Kepler's second and third laws are relatively small, and the approximate character of the initial model is also supported by the knowledge that Kepler's laws had good empirical support. The approximate character of the initial model lends Newton's conclusions some initial credibility, enough to warrant the effort to undertake the synthetic task of developing more realistic models that will be more adequate to known features of the solar system and will more fully incorporate Kepler's and Newton's principles. (Newton, 1962, Book III. See Glymour, 1980, for more details on these inferences.)

The inferences just described are drawn from what is likely to be a retrospective account in the third edition of the *Principia* (1725). In the absence of detailed documents regarding the actual course of Newton's reasoning, this account still can serve as a retrospective reconstruction of a strategy of reasoning from phenomena to theory that might be of heuristic value in other contexts. In short, it can be considered an account of the 'discoverability' of celestial gravity from empirical laws and mechanical principles that Newton has taken as established. There are two prominent components in Newton's procedure: By analysis Newton considers various components of the solar system in isolation, e.g. the earth-moon system, the system of Jupiter and its moons, and a system consisting of the Sun and one planet. These systems are the subject of his initial models within which the above inferences are conducted. The second component is the inductive extension, by appeal to Rule II, from common phenomena observed in these components, viz. circular motions obeying Kepler's second and third laws, to common causes of these phenomena, inverse square centripetal forces. A third strategy is the appeal to continuity between seemingly diverse phenomena, the mathematical continuity between the mechanical principles for projectile and satellite motion, and the inference of a common cause, inverse square forces, that act in different circumstances in these two extreme cases.

Newton's procedures thus described also have considerable heuristic merit, particularly if compared to, e.g., Popper's heuristic of free conjecture. Newton is able to cite powerful if not conclusive reasons for ascribing inverse square centripetal forces to the Sun and the planets and for identifying them with terrestrial gravity. Without this reasoning forces constraining planets could be directed anywhere and could be variously dependent upon spatiotemporal relations and other features of the planets. Inventing and trying mechanical models based on these various unconstrained possibilities would be a hopeless task.

Newton's Rule III includes a second strategy of combining analysis and induction (McGuire, 1970). By dividing observable bodies with extensive quantities such as figure, motion, and mass, one can bring before the mind or produce in actuality smaller bodies having the same attributes. The process of division may lead to the consideration of bodies that are too small to be seen with current implements, e.g. sub-microscopic components of observable objects. In this case, unlike inductions to celestial bodies, fully unobservable objects are considered. However the procedure of division permits something to be inferred about them, that they have the same mechanical attributes that observable objects have. If these components have the same attributes, perhaps the same laws apply to them, viz. the laws of mechanics, gravity, and electric and magnetic attraction and repulsion. This application of the analytic-inductive heuristic is implemented in speculations about light, heat and chemical affinity in Queries 29–31 in the fourth edition of the *Opticks* (1730). In these cases Newton takes an additional step of postulating special attractive forces for refraction, diffraction and affinity which differ from the macroscopic forces of gravity, magnetism and electricity. However atomic mechanical models of light interacting with matter and chemical reactions could provide conditions for deducing causes, the forms of the forces acting, from effects, the atomic configurations and motions that constitute the macroscopic objects thus analyzed.

Newton's 'Rules of Reasoning' are the basis of an inductive heuristic that can produce deeper theories than can the more phenomenalistic methods of enumeration and correlation. In the case of celestial gravity they permit the inference from observable effects to unobservable causes, and in the case of Newton's atomic speculations they warrant the extension of some constraints to the totally unobservable atomic domain. According to Rule III, whatever specific characteristics atoms have, they will include extension, figure, mass, the possibility of motion and they will obey the laws of motion. These constraints will place restrictions upon hypotheses worth considering in the construction of atomic models, from which Rule III obtains its heuristic merit. These are essentially the constraints that successful programs in molecular and atomic physics have operated under ever since. Unlike enumerative and correlative inductive heuristics, those of Rule III can be regarded as conceptual heuristics because they provide guidelines for conceptualization in the atomic domain.

Other scientific research programs have made successful applications of one or another aspect of Newton's analytic-inductive heuristics. In Geology James Hutton and Charles Lyell held that theories about the geological past should incorporate *vera causae*, that is 'true causes' or laws whose truth could be established as governing present day geological, climatological and biological processes. As in the case of planetary motions, the effects are observable and established laws are used to infer unobservable causes acting in the past. Accordingly, processes of mountain building, valley erosion, etc. are 'uniform' or slow but persistent in their action over long periods of time. It was argued that without this method nothing could be inferred about the prehistorical geological past (Kavoloski, 1974, Rudwick, 1969). In sum, the restriction of geological theories to those incorporating established laws or *vera causae* provided a heuristic. This heuristic is now considered to be limited by the belief that the conditions under which mechanical and thermal laws governing geological processes apply have evolved throughout the Earth's history.

In his 'Notebooks on Transmutation' (DeBeer, *et. al.*, 1960) Darwin applied much the same inductive procedure in inferences from observed processes of production for varieties domestic plants and animals to present and past causal processes producing species of organism in the wild. Plants can be observed in the garden to flourish in some soil, humidities, conditions of light and shade and competition from other plants and animals, but to perish or to suffer lowered fertility in other circumstances. Those of a given species also vary when they sexually reproduce under favorable circumstances (B3f). The principle that adaptation to circumstances determines survival and propagation, Darwin's principle of survival of the adapted, is extended to wild species, which, like domestics, depend upon their surroundings for nourishment and are subject to death or low fertility from predators, diseases and other environmental adversities (B38,84). Human individuals and families have the same dependencies for survival and successful reproduction, and humans must produce fertile offspring to be genetically represented in subsequent generations. Repugnance to marriage, infertility or death can prevent a family from being thus represented. Since wild plants and animals are equally subject to failure in fertilization, disease and premature death, these 'families', viz. reproductive lines or varieties, that are most fertile because of superior abilities to attract a mate, resist disease, and successfully propagate offspring, will be subsequently represented at the expense of those less well adapted. Also from human demographics Darwin observes that relatively few of the families prevailing earlier in history can be represented now, especially if a constant population is assumed and presuming the average family has more than two offspring. Some human families must inevitably suffer extinction, if not by death, at least by infertility. Applying this reasoning to species, Darwin concludes that species extinction is the inevitable consequence of species splitting, which might be called Darwin's principle of inevitable extinction. Thus species extinction, not the episodic extinction of whole genera by senescence, as Darwin originally thought, should be a regular occurrence in phylogenetic history.

After rereading Malthus' *Essay on the principle of Population* Darwin extended Malthus' principle of universal superfecundity to all living organisms, and applying his principle of inevitable extinction now to individuals of a species, he concludes that death of individuals within a species before they reproduce, and other sources of infertility as well, must prevail inevitably within the populations that constitute every species. From this conclusion and the premises that sexually reproductive organisms vary, and vary in their adaptedness, as established in the first part of Notebook 'B', and the principle of survival of the adapted, Darwin deduces the principle of natural selection: Everywhere in biological nature there is a continuing 'sorting out' of adapted characteristics and these characteristics are preserved because they favor the propagation of offspring and thus the survival of populations constituting a species.

Darwin thus extends concepts borrowed from observable genetic processes in domestic animals and plants and from observations in human demographics to the wild in an effort to discover the 'laws' governing natural speciation. The principles that he gets from this effort are (1) that species' probable survival and reproductivity are proportional and (2) that superfecundity in families and in species entails extinction in each. The effects of (3), the principle that adaptive variation occurs at birth at various rates are, by (1), survival and extinction. The specific causes of reproductive success and failure are various, but each can be established as *vera causae* from observations on domestic species and human beings. By a similar process of extension from Malthus' observations about human beings, Darwin (4) claims that all species, even the most slow to reproduce, tend to multiply geometrically thus causing universal and inevitable population pressure. Universal superfecundity is thus also a *vera causa*. Malthus' principle gives an antecedent for (2), a driving force producing pervasive extinction in nature. From these premises, Darwin deduces an effect, the sorting and propagation of adapted structure.

Darwin's heuristic for discovery is 'inductive' in the sense of Newton's 'Rules of Reasoning' and his project of deducing from the phenomena. Rule III permits the extension of laws over subjects which when divided can be described in the same terms, i.e. mechanical laws can apply to the components of material bodies because their components are equally describable in mechanical terms. Inferring characteristics of components from characteristics of compounds can be considered a variant of the method of analysis. Darwin was quite aware that propagation, extinction and survival are terms that apply to individuals, families, races or varieties, and to species as well. The reverse of Newton's method of analysis is composition, inferring from characteristics of component to those of the compound. When wholes are described in the same terms as their components, look for laws governing compounds that are analogous to those governing components. As individuals, domestic animals and plants, human beings and wild organisms of many species share in sexual reproduction, disease, nutritional needs and predation. Lyell had noted that extinction of species consists of the gradual extinction of the individuals that compose them, that is a gradual decrease of a population to nothing. The converse of this process would be the propagation of a species, which, according to Darwin's compositional heuristic, would consist in the successful propagation of individuals resulting in an increase of the population. Thus Darwin arrives at the principle of propagation of the adapted, or the 'survival of the fittest'.

A second procedure Darwin shares with Newton is one of extending *vera causae* over an established natural kind, a procedure that has been called 'transduction' (McGuire, 1970). For Newton this procedure allows him to extend, by reference to Rule II, according to which like causes should be sought for like effects, terrestrial laws of motion and gravitation to the celestial region. Similarly Darwin reasons that genetic laws governing one of the groups, humans or domestic animals and plants, should also govern the other, groups of wild organisms. Lyell's methods prescribe the extension of laws established as *vera causae* in time, and for Darwin this means that genetic laws found in humans and domestic organisms can be extended into the distant phylogenetic past.

Finally, the deduction of natural selection from 'phenomena', in this case empirical laws such as universal superfecundity, using principles of variation, inevitable extinction and survival, is comparable to Newton's deduction of uniform terrestrial gravitation from Galileo's distance theorem and inverse square gravitation from Kepler's laws. As in Newton's case, this inference is not purely logical, but requires a background of laws such as variation, extinction and survival.

Newton's inductive procedures can be considered heuristics for discovery in these senses: The *vera causa* inductions are grounds for initially formulating hypotheses in certain terms, viz. the same or similar terms that govern the formulation of hypotheses in known domains.

However, the deductions are inferences to the credibility of hypotheses whose terms and formulations are already available. Thus they provide initial evidence for hypotheses once they have been formulated by other means. Thus the Newtonian deductions might be put into the context of justification. But the *vera causa* inductions allow extensions of terms to unobservable domains and also endorse these extensions as inductively probable. They provide evidence for conceptual choices which, given that induction is the best procedure for extending our knowledge into the unobserved, are our best hopes for epistemic success (Reichenbach, 1938). Hence they also have a justificatory aspect. Thus these methods confirm the point that there should be no sharp separation between procedures for conceptualization or formulation and those for justification.

Darwin takes concepts of adaptation, reproductive survival, sexually induced variation and extinction from various sources. By a combination of inductive extensions of concepts and deductive affirmation he arrives at a conception of a cause for pervasive and universal picking of adapted structure, natural selection. Selection is hardly a new concept, for breeders have used it since neolithic times. Its application to natural species may have been initially suggested to Darwin by breeder's pamphlets that he read in the summer of 1838 (Ruse, 1975b). However, passages in Darwin's transmutation notebooks indicate that he held the received view that domestic varieties were artifacts that could exist only under human protection, not in the wild (C17e, 53, 219). Only Malthusian superfecundity made it plausible to apply this concept to nature, and this plausibility is what Darwin's deduction shows. Darwin's deduction is thus an essential part of the discovery of natural selection.

In this case the context of justification consists of efforts to further substantiate hypotheses whose initial formulation and support comes from Newtonian inductions and deductions. In Newton's case further justification comes from applications of the laws of motion and gravitation to one, two and several body models of planetary systems from which observable deductions can be drawn (Cohen, 1980, Lakatos, 1970). In the transmutation notebooks Darwin tried to test various induced genetic laws governing variation and inheritance against reports elicited from breeders, and he tested genetic hypotheses about relations within biogeographical genera against biogeographical reports from travelers. In the *Origin* he reports testing other components of his theory, particularly the 'principle of divergence', according to which a multiplicity of species is better adapted to the utilization of physical resources than are relatively few species. This principle was tested in experiments with plantings in his fields. Also he applied natural selection to reports of wolf varieties in the Catskill Mountains (1964, p. 90). Furthermore Darwin points out in the *Origin* that consilient evidence for evolution and natural selection can be found in biogeographically related genera, the occurrence of vestigial organs, the homologous structures described by comparative anatomists, and the similarities among embryos of various mammals and vertebrates. These can be considered justificatory arguments based upon applications of hypotheses generated by inductive extensions of concepts and hypotheses by division, composition and transduction.

The discovery of evolution by natural selection partially supports N. R. Hanson's views of the context of discovery, as expressed in these passages:

> What would be our reasons for accepting H? these will be those we might have for thinking H true. But the reasons for suggesting H originally, or for formulating H in one way rather than another, may not be those one requires before thinking H true. They are, rather, those reasons which make H a *plausible type of conjecture.*
>
> Other *kinds* of hypotheses were available to Kepler: for example that Mars' *color* is responsible for its high velocities, or that the dispositions of Jupiter's moons are responsible. But these would not have struck Kepler as capable of explaining such surprising phenomena. Indeed, he would have thought it unreasonable to develop such hypotheses at all, and would have argued thus. (1961)

A 'plausible type of conjecture' is one which, to the best of one's knowledge at a given stage of inquiry, has the best chance of surviving in the 'context of justification', that is, in the epistemic circumstances in which further explanatory models and tests are sought. In Newton's case the epistemic presumption of causal uniformity made celestial gravity and various atomic forces of attraction and repulsion plausible conjectures. They are so presumed because the *vera causae* for new domains can be presumed true as the best hope of describing and explaining phenomena in the new domain, provided there is no reason to the contrary (Rule IV). This extension is thus judged more likely to survive further experimental tests on the grounds that inductive empirical evidence initially supports it. As will become apparent shortly, Hanson's view of the 'logic of discovery' is not that transductive and deductive procedures should be included but that initial evidence for a hypotheses should be that it explains puzzling phenomena. Nevertheless, the above considerations should indicate that Newton's deductive and transductive meet Hanson's conditions for such a logic.

6.4. ANALOGICAL METHODS

William Whewell pointed out in reviewing Lyell's *Principles of Geology* in 1831 that causal hypotheses should not be restricted to *vera causae*. There should be no *a priori* assurance that the same causes now operative always have been so in the past (Ruse, 1979, Chapter 3, Kavaloski, 1974). Also at the time he wrote this criticism there was some geological evidence, e.g. recent fossils high in the Alps, that suggested to geologists such as Adam Sedgewick that the Alps were raised by causes far more violent than those observably acting today.

At one time to some these considerations appeared as applicable to theories of light as to geology. Newton's emission theory accords with his *vera causa* strategy insofar as light corpuscles are described in the terms recommended in Rule III. The effects of rectilinear propagation, reflection and attraction can be observed in the behavior of macroscopic projectiles. By contrast, as Newton conceived it, the 'ether' through which light pulses or waves are propagated at very large, if not infinite velocities, seemed not to be describable in terms applicable to observable phenomena, particularly since this propagation theory seemed to Newton to require infinite forces in luminous bodies, and there appeared to be no analog of water and sound waves passing around barriers in the behavior of light (1730, *Query 30*, Achinstein, 1987). That is, Cartesian pressure and wave theories of light (i) failed to meet the like effect condition of Rule II particularly concerning rectilinear propagation and (ii) implied causes of infinite or near infinite velocities of propagation that could not be observed.

However, both shortcomings were met by defenders of the wave theory, particularly Christian Huygens who incorporated Romer's observation in 1675 of the finite velocity of light into his wave theory. According to this theory light is the propagation of disturbances in a mechanical ether similar to the propagation of sound vibrations in air. Huygens' success in applying the theory to obstructing, reflecting and refracting surfaces for the deduction of laws governing these phenomena should have given the wave theory equal credibility with the emission theory. Light waves could as plausibly as emissions be argued to be *vera causae* from observations of water and sound waves. The theory was subsequently supported by empirically successful models of reflection, diffraction and refraction. As in the emission theory, the disturbances in the ether constituting light are described in the same terms as water and sound waves. Light waves consist of displacements of ether components that are propagated by impulse to neighboring regions. However Huygens did not take the next step suggested by the analogy with sound and water, viz. to seek a mechanical theory of ether vibrations by ascribing to the ether causes like those in sound, namely mechanical restoring forces proportional to the displacement of the ether components.

Hence suggestions by Laudan (1981) that the wave theory should, in contrast with the particle theory, be considered to be hypothetical and not initially inductively supported, or by Achinstein (1987) that inductive support of the wave theory is just by elimination, do not take account of the actual history of this theory.

Still Whewell's attack on Newtonian inductivism points to possible limitations of transductive strategies to domains throughout which certain kinds of uniformity can be expected or hoped for. Unobservable domains need not be describable in the same terms as those observed, much less contain events and processes subject to the same cause. That the procedures of Newton's Rule III can break down has long been known as fallacies of composition and division, where it is recognized that parts need not be described in the same terms as wholes and conversely. This breakdown is particularly evident in the history of physics in which it was eventually learned that macroscopic quantities, such as temperature and entropy, cannot be ascribed to all components, such as molecules and atoms, and that the mechanical quantities of location, momentum, energy and duration, which ideally have exact values in macroscopic application, do not have equally determinate values in all circumstances when applied to atomic and sub-atomic components. Similarly Newton's Rule II can fail when various causes can produce similar effects, as would be implicit in Newton's own acknowledgment of magnetic, electric as well as gravitational forces, all of which, by Law II, can have the same mechanical effects, viz. centripetal acceleration.

One might seek a compromise between, on the one hand, Whewell's more permissive willingness to include as hypotheses non-natural catastrophic causes, particularly in cases where the effect is biological organization, and, on the other, Lyell's and Darwin's procedures, by rejecting uniformitarianism but retaining actualism (Rudwick, 1969). In this case one would restrict geological inquiry to the igneous, aqueous and other agencies now observable, but would not be restricted by the assumption that they have acted with the same energy as now. In effect causal laws and conditions are to be the same kind as can be observed, but the circumstances in which the causal conditions occur, the boundary conditions in any particular instance, may differ in value from what can now be observed. The Alps could, then, have been produced by the same igneous forces now observable in volcanos and earthquakes but at a time in which these forces acted with greater frequency or intensity than at present.

This kind of compromise fits John Herschel's more liberal variant of transduction:

> If the analogy of two phenomena be very close and striking, while, at the same time, the cause of one is very obvious, it becomes scarcely possible to refuse to admit the action of an analogous cause in the other, though not so obvious in itself. (1830, 142)

"Analogy" is a rather elastic term, particularly if 'analogous' effects and causes are analyzed as happenings that are 'similar'. Presumably most happenings of interest to science, particularly to naturalists and biogeographers, have many attributes. Since it would be methodologically perverse to 'close the book' on the possible discovery of more such attributes, we can say that they are indefinite in number. Similar happenings are plausibly thought of as those that share some but not all of their attributes. Thus there are an indefinite number of ways in which happenings might be similar. It would then appear that the requirement that unobserved processes have properties 'analogous' to those that are observed leaves open an indefinite range of possible hypotheses. This lack of constraint on possible hypotheses is the basis of criticism frequently offered against the 'method of hypothesis', where it is suggested that 'anything goes' provided that it has, possibly when embedded among auxiliary beliefs, laws or hypotheses, consequences that are in some manner experimentally testable. Without further constraint regarding preferred hypotheses or preferred relations of analogy it appears that hypothetical and analog methods are equally heuristically inefficacious, and thus neither offer suitable 'logics' of discovery.

Herschel seems to be aware of this problem, as indicated in the following remarks:

> The classifications by which science is advanced, however, are widely different from those which serve as bases for artificial systems of nomenclature. They cross and intersect one another, as it were, in every possi-

> ble way, and have for their very aim to interweave all the objects of nature in a close and compact web of mutual relations and dependence. As soon, then, as any resemblance or analogy, any point of agreement whatever, is perceived between any two or more things, —be they what they will, whether objects, or phenomena, or laws, —they immediately and *ipso facto* constitute themselves into a group or class, which may become enlarged to any extent by the accession of such new objects, phenomena or laws, agreeing in the same point, as may come to be subsequently ascertained. It is thus that the materials of the world become grouped in natural families, such as chemistry furnishes examples of, in its various groups of acids, alkalies, sulphurets, &c... (1830, 134)

Thus the 'particularly striking analogies' that Herschel highlights are those based upon shared properties that cluster the objects of investigation into 'natural classes', viz. classes defined by the common possession of several unchanging characteristics, such as biological species, or several characteristics that differ only in degree, such as newtonian material bodies, and classes distinguished by 'polar' characteristics (1830, 135), as are acids and bases.

Qualitative likeness with difference only in degree is entirely compatible with Newton's 'like' causes and effects, for celestial bodies and atomic corpuscles both possess mechanical qualities but differ from ordinary bodies only in magnitude. Newton's constraints would also permit geological actualism without the uniformitarian assumption.

Presumably for extending causal relations beyond observable domains, these preferred classes should be partially defined by characteristics with causal efficacy bearing upon interactions between objects within one class or between objects in different classes. Thus mass, location, electric and magnetic charge distinguish material bodies as natural classes because these quantities are also causally efficacious in explaining motions and trajectories. They are preferred or 'intelligible' attributes because in their terms laws of motion can be formulated. 'Like' or 'analogous' effects and causes thus can be those that are described in these same terms and by the same laws that vindicate the choice of these terms, as Newton suggests in Rules II and III.

Mary Hesse's 'Real analogies' and 'experimental identifications' (1974 , p. 266) include strategies relevant to the identification and extension of natural classes. As a means to such identification one might seek to identify two apparently distinct properties or entities and to establish that they are interchangeable in different contexts, wherein the apparent differences emerge. Thus static and current electricity, electromotive force and static potential are respectively identified, and conduction currents and displacements are respectively effects of the application of this potential across respectively conductors and insulators. This project of identifying the two kinds of electricity and conducting investigations into the consequences of this identification was initiated by Michael Faraday in the 1820's.

This procedure for identification depends upon the general causal conception that one law, when applied to the same or similar entities under different circumstances will generate different effects, a conception exemplified in the well known application of Newton's laws to a projectile in various initial conditions to give parabolic, elliptical or hyperbolic trajectories. However, the details of the causal process Maxwell proposed are derived from the application of this conception to more specific spatiotemporal and causal configurations, viz. the 'field' exemplified by fluid mechanics. A similar procedure leads to the identification of electromagnetic waves and light. In fluid mechanics waves can be described as displacements under linear restoring forces that are propagated through the medium with velocities dependent upon mass-densities and elastic restoring forces. Correspondingly an electromagnetic wave can be construed as an exchange of electric and magnetic energies, corresponding respectively to mechanical potential and kinetic energies in fluid media, and, as in fluids, this wave will be propagated with a velocity dependent upon dielectric and magnetic coefficients of a given electromagnetic medium. An essential step in Maxwell's procedure here is the extension of electrostatic and magnetic capacities to free space, a move that he made as a means to obtaining a source-free electric field comparable to Faraday's magnetic fields. As magnetic bodies concentrate lines of magnetic field intensity, so there should be source-free lines of 'displacement' permeating free space and becoming concentrated polarized matter. This identification of natural kinds, viz. the electric and magnetic 'stresses' within and outside of matter, allowed him to infer the applicability of Faraday's laws of electric and magnetic induction to free space.

It then is from a wave equation set up in terms of these field quantities and from theoretical links between electrostatic and electromagnetic units of charge from his theory of electric current that Maxwell computes the velocity of an electromagnetic wave in free space, and, as is well known, this value coincides with the empirical value for the velocity of light. This coincidence of one essential property provided initial evidence that light is an electromagnetic disturbance and that the laws applying to one, e.g. laws governing electric and magnetic fields in matter, should also be observed in the other, i.e. they should permit the computation of correct values for various indices of refraction.

This method of 'real analogy', however, is only one heuristic among several for generating 'experimental identifications' or empirical grounds for extending natural classes. In some cases the apparent differences that distinguish classes mask underlying and more essential similarities not because one law is being applied in different circumstances but because inquirers may initially be unaware of common causal processes essential to the existence of the elements of the distinguished classes. Thus Darwin realized that the fundamental process for explaining the existence of any living organism, whether human, domestic or wild, is reproduction, whether sexual or a-sexual. Accordingly one should expect similar laws governing the production of humans, as well as wild and domestic animals and plants, regardless of apparently distinguishing 'moral', sensory and vegetative capabilities. Once this natural class is identified even these distinguishing characteristics can be seen as the same kinds of characteristic, viz. adaptations or means by which their occurrence is preserved in the reproductive process.

In this case the strategy is to find one fundamental causal process among apparently diverse classes and to identify them as members of one natural class on the assumption that causal processes are more fundamental to picking natural classes than are outward appearances. The next step is to show that these different appearances contribute in the same way to the operation of the causal process by which the natural class was initially identified.

This identification then sets the stage for *vera causa* inductions. In Maxwell's case, an important inference was from laws governing magnetic interactions of conduction currents to laws governing like interactions in displacement currents. In Darwin's case some of the several important inferences from domestics and humans to wild animals and plants have already been described.

Also Herschel's suggestion that 'laws' can be compared and classified indicates that he might allow what Maxwell called 'physical analogies' in his inferences to *vera causae*. Physical analogies consist of "...that partial similarity between the laws of one science and those of another which makes each of them illustrate the other... (quoted in Hesse, 1974, p. 261). They do not require that the phenomena compared be described in the same terms, as does Newton's 'likeness' of cause and effect. Coulomb's law of electric attraction is physically analogous to Newton's law of gravitation in its distribution of force through space, though it falls short of identity with gravity because of its origins in charges rather than masses and it includes the possibility of electric repulsion as well as attraction. Still Coulomb's law would have 'inductive' support in Herschel's methodology. Herschelian methods would also justify the above-mentioned suggestion that the electric stress in a medium should be conceived to be source free, as is magnetic field intensity. In this case distinct terms, magnetic field intensity and electric displacement, are given common spatiotemporal configurations, inferences are drawn to like laws governing the distribution of these fields.

These physical analogies are particularly heuristically powerful when there is a mathematical formalism and a set of mathematical techniques or heuristics for drawing consequences of particular distributions of quantities or for solving problems involving the application of physically analogous systems of description to particular cases. Accordingly, inverse square central forces, whether electrical or gravitational, produce parabolic, elliptical, or hyperbolic orbits under analogous initial conditions, a physical analogy exploited in the early stages of the development of the nuclear theory of the atom. Vector and scalar field theories have methods for dealing with fields originating and terminating in sources and sinks, for rotations or vortices, and for the propagation of waves, which were originally developed for the mechanics of fluids and theories of sound, but then were applied by Lord Kelvin and Maxwell in drawing consequences of Faraday's laws for the behavior of electric and magnetic fields.

In sum, these analogical heuristics emerge from the examples just discussed:

1. Identities between entities or common membership in a natural kind can be established on these grounds:

a. The entities have like properties that are, under a system of background beliefs B or a law-concept complex C, essential to their identification or causally linked to their occurrence or behavior.

Thus the velocity c and transverse vibrations are essential to the existence and propagation of light. Also electromagnetic vibrations are transverse and must be propagated at c in free space. It follows that probably light is an electromagnetic vibration. Maxwell anticipates tests of the hypothesis thus generated by drawing its consequences for indices of refraction in various insulating materials and in diamagnetic and paramagnetic media. Similarly, by reference to reproduction, susceptibility to disease, ecological and behavioral similarities, Darwin identifies humans and non-humans as members of a common natural kind.

b. To strengthen these 'real analogies' one can show that apparent differences are either importantly the same kind of attribute common to all in a given natural kind or that they emerge as causal consequences of one or several things having different circumstances or different kinds of organization.

Thus increased intelligence in humans is just greater adaptive ramifications of the perceptual powers, the instincts and the habits found in non human species. The magnetic effects of charges are the consequence of charges being in acceleration, and hence these effects emerge only in circumstances when static or moving charges are subject to acceleration.

2. Once 'real analogies', identity or common membership in a natural kind are established, inductive presumptions that laws manifest in one member can be extended to others on the grounds that common causally essential properties are good indications that common laws hold.

Laws governing electric and magnetic behavior in various materials should hold or have specific consequences for light. Laws governing humans should govern non-humans, and conversely.

3. Once common membership in a natural kind is established, concepts known applicable to some members may be extended to others, or other concepts may be found that can thus be extended.

Human families may fail to propagate because failure to find a mate, premature death from various causes such as accident or disease, or other causes of infertility. Conversely, attractiveness to the opposite sex, resistance to disease, and other causes of fertility can explain fecundity which in turn makes propagation more probable. The same causes of propagation or failure to do so can be extended to races and species, which are just larger and more distantly related families.

Magnetism in material bodies is construed as polarization of their small parts, as can be inferred by Newtonian division. These characteristics of magnetic substance are, by Faraday's conceptualization, equivalent to the concentration of lines representing magnetic field intensity and the initiation and termination of lines of magnetic force. The lines of magnetic field intensity represent magnetic conditions in matter and in free space, and thus one natural kind, a magnetic field, can exist with or without the presence of matter. In electrified bodies polarization also occurs between oppositely charged components of molecules. As in magnetic fields in free space, the presence and various directions of electric forces can be observed in free space. As in the case of magnetic dipoles, the presence of electric dipoles can be represented by lines of electric 'displacement' which never terminate but become concentrated in electrified bodies. Thus concepts initially applied to matter, viz. polarity, are represented as concentrations of lines measuring the intensity of some agency so that the agencies they represent can be extended to free space, viz. as continuations of these non terminating lines. Laws governing the interactions between these agencies, as observed with polarizations and intensities in matter, may also be found to apply where these agencies exist outside of matter.

These analogical heuristics are attempts to give the similarities observed between two apparently distinct objects, populations, or systems sufficient import that there is some rational warrant for transferring concepts and laws known or observed in one to the other. The 'uniformity of nature' that the inductivist hopes for need not be the applicability of identical concepts and laws throughout the entire universe, as Newton hoped for the universality of mechanical concepts and laws. Though Maxwell held mechanistic hopes similar to Newton's, his and Faraday's procedures of conceptualization did not restrict them to mechanical concepts for electric and magnetic fields. The distribution of these fields under various circumstances and their interactions in electromagnetic induction were observed and formulated in qualitative laws by Faraday without any essential speculative hypotheses as to their possible mechanical composition. Thus qualitative laws governing the distribution of magnetic fields about conduction currents as well as natural magnets, the induction of conduction currents from other such currents and from changing magnetic fields, and the distribution of static electric fields about various charged material bodies were drawn from experiment in Faraday's researches. The concepts used to express these laws, viz. force or flow intensities, their concentration or dilation in space, their sources or sinks or the absence thereof, and their rotations, are formed with spatiotemporal and dynamical configurations, and these configurations are borrowed from fluid kinematics. Fluid kinematics provided a spatiotemporal and dynamic framework within which to fit concepts sought for describing and explaining the distributions and interactions of forces, fields and currents observed in the electromagnetic experiments. The mathematical formulations of fluid kinematics already contained the conceptualization needed for quantitative expression of the elements of this framework, and thus provided further constraints and positive procedures to aid Maxwell's search for quantitative electromagnetic laws. Specifically, Maxwell sought vector and scalar quantities that fit Faraday's field concepts and that could be dealt with by established mathematical procedures. Before its completion and subjection to further test, the credibility of Maxwell's project was based largely upon hoped for links between the behavior of electric and magnetic phenomena within and outside of matter similar to the linkage between fluid statics and dynamics. Maxwell also hoped for uniformity in the sense that the same laws should govern the behavior of electromagnetic agents when acting within and outside of matter. The first uniformity need not have entailed an identity of mechanical and electromagnetic terminology, just an identity of part of this terminology, viz. the spatiotemproal distribution and behavior of entities and forces which are more specifically defined in different terms respectively for mechanical and electromagnetic applications.

It does seem a lot less plausible to presume *a priori* that these more abstract uniformities would link fluid mechanics with electrodynamics, than that these domains should be the same both in terms of their abstract dynamics and particular state descriptions. However, this lack of initial plausibility may reflect current background ontologies B which distinguish electromagnetic fields from any material configurations or motions as separate natural kinds. In Maxwell's case Newtonian mechanism was still a credible background ontology. Although the mechanical ether was problematic, he had no decisive reason at the time for rejecting it. Though Maxwell eventually rejected the specific fluid analogies used in various papers in which his field theory is presented, he remained committed to the view that the ether is mechanical and is governed by energetic laws, viz. Lagrangian mechanical principles. In the context of a mechanistic ontology fluid mechanics would seem to be the natural conceptual system to try to apply to continuously distributed fields inside and outside matter. Within that context there is a presumption of terminological as well as spatiotemporal, dynamic and energetic uniformity linking material fluids and electromagnetic entities and processes. Hence, in Maxwell's conceptual context, his procedures had epistemic support. They were not based on pragmatic hopes that new phenomena could be understood in the same or similar terms as those already successfully in use just because the latter were all that were available, or just because they were all that were sufficiently developed to provide the required quantitatively exact and deductively rich descriptions.

One need not be committed to a mechanical ether to have some reason for projecting established laws to electromagnetic domains if the laws under consideration are abstract principles such as energy conservation, entropy increase or the principle of least action. Energy, entropy, momentum and action can be realized in mechanical terms in many configurations, whether distributions of force and motion in a variety of particle configurations or distributions of motion and pressure in a field. They can also be realized in a variety of non mechanical terms, as in various distributions of electric and magnetic field intensity, distributions of frequencies and intensities in a radiation field linked with material oscillators, and so on. Maxwell's procedure did not commit him to one particular mechanical conception of the electromagnetic ether, that is one mechanical model embodying the distribution of these quantities in the electromagnetic field. He used various plausible models on various occasions in his inquiries, and, though he never abandoned his mechanistic ontology, he also never finally endorsed a specific mechanical model of the ether. His final position is entirely compatible with a non-mechanical model: The electromagnetic field in free space embodies energy, action, entropy and momentum in non-mechanical terms. Hence the method of physical analogies is not constrained in the way that Newton's *vera causa* procedure is, that is to the same fundamental descriptive terminology for observed and unobserved domains.

A procedure similar to Maxwell's was used at the turn of the

20th century in attempts to theorize about atoms and radiation. Again, abstract principles such as Lagrangian and Hamiltonian energetic principles and the laws of thermodynamics were presumed to hold for atoms and radiation fields and descriptive terms were sought that would provide empirically satisfactory models of these entities and their interactions. In these developments there was a division among the prominent investigators in the terminology chosen for the models. Some, such as Heisenberg, initially preferred to compute the magnitudes or frequencies and the probabilities or amplitudes of changes in energies without constructing any mechanical models of the atom. Others, such as Schrodinger, sought models in terms of wave mechanics, where the traditional mechanical concepts of location, energy and momentum were defined in terms of the wave concepts of superposition, interference, frequency and wavelength. These wave concepts shared at least some general dynamic and spatiotemporal features with mechanical and electromagnetic waves, though their substantive terms could not be identified with those in the classical waves because of their dimensionality and their use of imaginary variables. Still others, such as Bohr, thought that traditional mechanical terms could be applied only in special experimentally realizable circumstances and that both wave and particle terminology is required for models of atomic and radiation processes. Finally, Einstein defended the position that there should be no restriction on the kinds of terms used to describe radiation fields and atoms and that physicists should hope for deterministic laws in terms that can be radically different from those that describe the macroscopic processes. Still his researches began with the extension of established principles, such as the conservation laws, invariance principles and general covariance, to new domains of inquiry.

6.5. THE METHOD OF HYPOTHESIS AND ABDUCTION

The 'hypothetico-deductive method' (hereafter referred to as HD) is often thought of as a method for testing hypotheses or theories whose initial formulation is already in hand. It is said to be a procedure appropriate for the context of justification, not for the context of discovery. However, this view leads to a puzzle regarding Popper's choice of titles for his *Logic of Scientific Discovery*, or the 'logic of scientific inquiry', which is sometimes offered as a better English translation of the original German title. As suggested in Chapter 1, Popper's choice of titles is not so puzzling if discovery is thought of broadly as the production of novel knowledge, including novel concept, proposition, evidential relation and actual empirical evidence. Popper's views can plausibly be construed as reactions to the restrictions placed by inductivists and positivists on the introduction of novel concepts or the 'construction' of scientific theories in the conduct of scientific research. The strategy Popper recommends is not to restrict conceptualization to terms and laws that apply to observable domains or terms that can be logically constructed from such observables. The only logical restriction on new terms beyond their mutual consistency and compatibility with background knowledge is that they have maximal empirical content. This requirement is assured if with the new terms is also conceived as large as possible a number of circumstances that can entail the falsity of the hypotheses formulated in these terms. This last clause can be said to prescribe the search for a certain kind of evidential relation R.

The introduction of a new concept, such as natural selection, and the conception of possible circumstances that would entail the falsity of hypotheses using this concept, such as the appearance of traits exclusively to the advantage of species other than the one considered to be selected, are two independent acts. The first can be performed without the second, and conversely. As Feyerabend suggests in another context, various components of a scientific belief system and the technical powers of scientists to carry out empirical programs are logically independent of one another and therefore can develop independently and can have independently varying degrees of credibility or value as evidence for truth. Thus the beliefs that connect a theoretical term to empirical concepts, e.g. dynamical principles whereby the effects of motion can be empirically determined, can be logically independent of those that implicitly define a term, in this case the kinematic principles for describing location and motion. One set can be more or less epistemically adequate than the other. A concept thus need not logically entail knowledge of how it is to be tested, for kinematic principles for describing motion are logically independent of dynamical principles describing the manner in which motion is caused and the effects of motion on observers or other bodies. If there is a logical relation between a newly invented concept and observable circumstances, it also may be discovered after and thus independently of the initial formulation of the hypothesis. This is because knowing something cannot entail knowing all of its logical consequences, as pointed out earlier. In the case of Einstein's equivalence principle, which can be said to implicitly define the kinematic conception of gravity, its implications for light require the consideration of independent optical principles and its implications for planetary motion require independent consideration of circumstances prevailing near a large mass. Whether either of these can be considered 'logical' relations to empirically accessible circumstances, Einstein's discoveries of the principle and of its testability are sequential and not simultaneous, and should therefore be considered two independent episodes. Thus it is appropriate to allow 'breathing space' wherein an investigator can evaluate the promise of a newly invented concept and then seek novel or better means of establishing its testability, whether or not this testing should be 'severe' or directed deliberately to possible falsifying circumstances.

Feyerabend's position that an inquirer should be given breathing space to discover the testability of a newly invented concept thus removes all of Popper's 'logical' restrictions on new hypotheses or concepts. Thus we arrive at Feyerabend's 'methodological anarchy' in the context of discovery.

However, this position is the equivalent of having no guidelines whatever in concept formation. For a given problem there is no indication of preferable choices, or no prior indication of which, among the large number of possible conceptions that a free ranging mind might hit upon, warrants efforts toward discovering means of testing. This null heuristic has an advantage of no bias or prejudgment as to what concepts or hypotheses are epistemically promising or worth working on. However, discussion in earlier chapters shows that it is doubtful that scientific research ever does or can operate without any prior guiding conception of a subject of inquiry, whether of the kind of terms needed for its description of the kind of laws it should obey, or of relations it has to other subject matters. The very formulation and rational acceptance of a problem is the acceptance of a bias, viz. certain beliefs that define a subject matter and an epistemic objective. These beliefs provide a framework of heuristic import insofar as further conceptual or empirical research must be compatible with all or most of it.

Rejecting a-heuristic research is not denying that the assumptions or biases underlying heuristics, including the formulation and choice of a problem, are fallible or limited in their applicability, or that there are a number of different kinds of guideline that might be appropriate in different circumstances. It is just that it is unrealistic to suppose that scientific inquiries ever operate without guidelines. Suggestions that 'playing' with equations (Feynman, 1965) is a good way to make discoveries should not be considered to be the endorsement of entirely a-heuristic or anarchistic inquiries. Dirac is said to have arrived at a theory of the electron by playing with the Klein-Gordon equation, an energy equation for a relativistic classical particle. This enterprise produced a novel way of factoring the equation into linear components needed for his transformation theory. The discovery of the novel factorization may have been fortuitous, but it does not follow that Dirac's project was without heuristic direction. The Klein-Gordon equation represents the energy of a classical free particle in relativistic terms, and its application to free electrons is as much an inductive procedure of extension or generalization as is Maxwell's extension of energy conservation to the electromagnetic field. Dirac also used standard operator substitutions for energy and momentum to get his wave equation, and this substitution procedure was by then a standard heuristic for generating wave equations from classical energetic relations. It is based upon relations between energetic and dynamic formulations of mechanical laws established for classical laws and extended to the quantum domain, as in formulating Schrodinger's equations. Dirac also used an established heuristic that appeals to a 'correspondence principle' to link relativistic with known non-relativistic accounts of a free particle, which was a limited procedure for checking proposed theses in relativity theory against newtonian counterparts (MacKinnon, 1982, pp. 282f). Again, this procedure was worked out initially for laws using classical kinematic terms and then extended to the quantum domain.

As just suggested, the principal heuristic shortcoming of hypothetical methods is the absence of any restrictions on the concepts to be tried. Hanson (1958) was aware of this and suggested (see the quotation in the preceding section) that the plethora of 'silly' hypotheses and concepts permitted by a-heuristic variants of the HD method can be excluded by requiring that hypotheses and concepts can be sorted into promising and unpromising categories by an initial assessment of their explanatory power.

Hanson makes two points that can be summarized thus: (1) The heuristic permissiveness of the HD method can be restricted by requiring that novel conceptualizations exhibit their explanatory power before they are seriously considered. (2) There is a mode of appraisal distinct from acceptance or rejection, viz. entertaining a hypothesis or employing certain terms in its formulation, for which there is also a distinctive kind of evidence, viz. its apparent explanatory power. This kind of evidence could warrant a certain rational choice in scientific research, that is the choice of formulation H over H' as the one for which testable applications can then be sought. If H has greater explanatory power than H', then programs of research for H should be given the breathing space needed for developing or improving the auxiliary hypotheses and the technology needed for testing.

Hanson's position gives rise to a number of questions: (i) Is explanatory power the only ground for entertaining a hypothesis or employing certain concepts in its formulation? Might there be other kinds of evidence for initial plausibility? (ii) If a concept or formulation requires some development to determine its explanatory power, how do we decide which concepts deserve such development? This question concerns the heuristic value of hypothetical methods.

Problem (ii) is particularly evident in the deductivist account of explanation, where deductions presuppose quantitative or especially deductively fruitful qualitative expressions for concepts and formulations of hypotheses. In general, non-ad hoc explanations should appeal to laws which in different circumstances give different outcomes, and these distinct applications are items that may require extended research for their discovery. The search for such expressions thus can be a project requiring breathing space. But then is there any criterion for choosing those concepts most worthy of the breathing space needed for their development?

Darwin's researches in 1837–9, viz. those recorded in his 'Notebooks on the Transmutation of Species' (NTS) (DeBeer, 1960), are a project whose aim is to work out the various possible applications of an idea as well as to seek an idea that can be thus worked out. His project is to find genetic concepts whereby laws, and particularly *vera causae* can be formulated that will account for related species on various islands and in various continental geographical regions. In this case the search for applications include projects of determining conditions in various localities, biogeographical barriers that at various times separated populations in these localities, and specific adaptations that eventually distinguished emerging species.

In short, determining the explanatory power of a hypothesis seems to require some of the same kind of development that testing a hypothesis requires, at least short of the technological development required for gaining empirical access to certain applications. These considerations raise doubts about the heuristic import of Hanson's abductive procedures.

(iii) A third question raised by Hanson's position is whether there is a mode of appraisal distinct from epistemic appraisal or judging the believability of a hypothesis, viz. the appraisal of the worthiness of a concept or hypothesis to be worked out? There are two parts to this question: (a) Is there a mode of appraisal of laws, concepts and hypotheses which is distinct from acceptance with varying degrees of assurance? (b) Are the criteria for this other mode of appraisal distinct from epistemic criteria? It seems that a good case can be made for an affirmative answer to (a). One may work upon several different hypotheses without believing any of them in the hope that one might, with sufficient development and testing, either be believable or instrumental in determining what is believable, e.g. by turning out to be false. One may work upon competing hypotheses, and some (e.g. Chamberlin, 1891, Popper, 1958, Feyerabend, 1962, 1974) have held that exploring mutually exclusive and even radically different hypotheses is a good heuristic. However, one can hardly consistently accept a set of mutually exclusive hypotheses. Thus initial heuristic appraisal as distinct from decisions to believe seems well established.

However, it is far less plausible to hold that criteria for heuristic appraisal are qualitatively or more than quantitatively or circumstantially distinct from criteria for belief, particularly for the enterprise of pure science. It seems far more plausible, particularly if scientific research is considered an activity with epistemic objectives and a search for optimal means for achieving these objectives, to hold that criteria for heuristic appraisal are those that indicate a promise or potential that criteria for truth will be fulfilled. Such promise is most plausibly given by the partial fulfillment of epistemic demands for reliable induction, empirical consilience and explanatory power. Furthermore there seems to be no particular order in which these demands must be fulfilled. A hypothesis or theory may show promise initially either on inductive or abductive grounds, or it may prove to be empirically consilient before it is sufficiently well understood to be considered to provide satisfactory explanations or without being an analogical extension of a familiar law or theory.

Question (iii) can be put in more concrete terms: Is explanatory power evidence, albeit hardly conclusive, for the truth or believability of a hypothesis? Some philosophers, most recently Achinstein (1987a,b) have denied that explanatory power is grounds for likely truth. Achinstein's argument appeals to the point that everyday actions can have various good explanations, without there being any decisive evidence for the truth of any of these various hypotheses. Being paid a large sum of money can very well, if true, explain my writing this book, as can being required to do so for promotion, etc., but some of these good explanations are not likely to be true. Rather than their explanatory power, evidence for these hypotheses must be either direct, e.g. testimony regarding my predominating motives, or indirect via some background belief about my circumstances or general circumstances bearing upon philosophical writing projects, combined with some theory of normal human motivation. A consequence of this position, which Achinstein is willing to accept, is that successful explanatory applications of a hypothesis or theory, even one with prior credibility, does not further enhance its credibility. However, this consequence seems very much at odds with arguments that seem common and influential in scientific practice, e.g. those Darwin gives in the *Origin* to the truth of their hypotheses based upon alleged superior explanatory power.

Some other considerations still seem to weigh against the epistemic import of explanatory power: Darwin's hope for a naturalistic explanation of the biogeographical distribution of species in a genus is that there are non-ad hoc laws that apply in this domain and can be instantiated in observable domains, and that these laws also apply equally in the past, to humans and to wild and domestic species. Inductive arguments have been given epistemic import. They can provide at least initial evidence for believing, on the grounds that in the absence of other knowledge, particularly knowledge to the contrary, the best rational hope (Reichenbach, 1938) for knowing some unobserved domain is that it is like observed domains in some significant respect. For example, the same descriptive terms and the same laws, or just the same laws apply in the two domains. Genuine laws have counterfactual applications and thus are non-ad hoc in that they can be applied to different circumstances giving different outcomes. Hence, it seems that hopes for explanations in hypothetical terms are epistemically warranted just when the more liberal Herschelian inductions are warranted, that is when there is some warrant at least for extending established laws into the unobserved domain. But hypotheses or concepts that are introduced without this heuristic warrant also have no initial support because apart from the extension of laws established elsewhere, there is no reason for thinking that the new domain has the causal or ontic structure manifest in non-ad hoc explanations. That is, if there is no reason to believe that some new domain of inquiry can be described in the same terms as some domain known already (as stipulated in Newton's Rule III), or that laws governing the new are the same as those already empirically established in another domain (Herschel's more liberal requirement as illustrated by Maxwell's inquiries), then putative explanatory hypotheses in the new domain are not believable just on the basis of their explanatory power.

Perhaps this point can be made clearer by a little historical fiction: Darwin can be imagined to have 'played' with some ideas and hypotheses for which there was no initial warrant. Suppose some concepts C,C′,... gives a system of non-ad hoc laws L,L′,... and that these laws are applied to a variety thought experiments, e.g. about the possible split of a population of wolves into long and short legged varieties, the split of a population of birds into varieties with various kinds of beaks adapted to various feeding habits, etc. These thought experiments can demonstrate the explanatory power of the law-concept complex C,C′..., L,L′,..., apart from any evidence that the conditions imagined exist or that they actually produce any adaptive splitting of an initial population. This conceptual exercise can give good reason to hope that biogeographical ecology actually has this kind of lawlike organization, a causal organization that Darwin recognized as similar to the dynamics of celestial bodies. It may also lead to the discovery of means of establishing this kind of organization in this domain by leading Darwin to the kind of evidence that needs to be sought. However, without any reason for believing L,L′,... are true in some observable domain, or that canine predators ever live in open and scrubby country and have long and short legged varieties, there is no reason to believe that biogeographical ecology has this lawlike structure. With these thought experiments selection can be shown, if true, to explain various possible distributions of species, but such hypothetical explanations do not provide any evidence for their truth. Many naturalists and philosophers in the nineteenth century believed that the domain of living organisms contains phenomena that are causally ordered differently from the domains of physics, chemistry, and even geology and climatology. Rather than non-ad hoc applications of natural laws, they sought implementations of divine benevolent will in various geographical circumstances.

However, in the NTS Darwin appealed not only to the patterns of explanation found in astronomy, but also the kinds of explanation offered in Lyellian biogeography in defending his objectives for explaining relations among spatiotemporally proximate species (NTS, B100ff). Lyell's biogeography is an effort to explain the present distribution of animals and plants by reproductive and migratory powers, by the occurrence of species extinctions from natural causes, and by the disposition of various biogeographical and ecological barriers to migration. Hence Lyell's program dealt with the same subject matter as Darwin's own, the explanation of the present distribution of plants and animals throughout the world. Lyell's own program of explanation by migration and extinction establishes that the domain is lawlike in some respects, in respect to causes bearing upon migration and extinction, though it does not necessarily follow that it is lawlike in all respects, in respect to the production of new species. These considerations alone do not have the epistemic import that is afforded by the Herschelian method, which gives credibility to hypotheses in the new domain provided they are applications of laws already extended to that domain. With the information thus far specified in this paragraph, there is no general law governing regularities in migration and extinction that would support the hope that comparable regularities exist in the production of new species. This point, however, does not necessarily exclude some further epistemic considerations that will show that if a domain is lawlike in one respect, it is likely to be lawlike in other respects as well.

Darwin can and does make a stronger case for laws of speciation than just the premises that there are laws governing migration and extinction. Darwin's first premise is that the presence of mobile species over a wide range, including remote places such as the Galapagos Islands, is, as Lyell demands, naturally explained by their powers of locomotion and the geographical circumstances within their range. But secondly, varieties and species of a less mobile genus are often exclusively present in spatiotemporally proximate habitats. Furthermore, the older their habitat or the barriers separating their habitats, the greater the differences among these representative varieties and species. Also, it has been acknowledged since the work of Linnaeus that local varieties can be produced by natural means, that is, causes in their production are lawlike. Given the observable continuity between local varieties and local species, i.e. likeness in the effects of geographical or ecological separation, it would be arbitrary to insist that the local species had been specially created but the local varieties had colonized by migration and varied by natural means.

It is thus that Darwin offers inductive evidence that one aspect of biogeography, the production of species, is as lawful in its causal structure as is another, the spread of mobile species and the production of local varieties. The thought experiment described a few paragraphs above provides no evidential link of a proposed causal architecture to empirically supported beliefs, but this latter induction does link the lawfulness of speciation with the already empirically established lawfulness of locomotory and ecological agents. Lyell has already established that the biogeographical domain has some lawlike architecture. Darwin argues that the only ground for denying lawful architecture to the production of species requires an arbitrary distinction between local varieties and local species. Therefore there is good reason to believe that one further aspect of this domain is lawlike.

Also, under these circumstances, suppose Darwin comes up with a system of concepts and laws C,. . .,L,. . ., e.g. a concept of species defined in terms of geographical, behavioral, and other barriers to crossing, and a concept of species change by the acquisition of heritable variations. Also he offers laws governing variation and the transmission of variant characters under various ecological circumstances, and this complex explains putative biogeographical phenomena, whether actual or fictional thought experiments. The presumed lawlike character of the biogeographical domain would then give some initial credibility to C,. . .,L,. . . . There is reason to believe that the domain is generally lawlike in its architecture, and C,. . .,L,. . . instantiates this lawlike architecture. Therefore there is initial evidence for C,. . .,L,. . . . Thus Darwin seems to be correct in his efforts to defend his theories by their explanatory power.

Empirical evidence for Darwin's explanatory scheme can equally come from the empirical actuality of one of its applications, e.g. the various species of South American camelids, the distribution of South American mockingbirds, and so on. Furthermore, if components of the law-concept complex C,...,L,... can be established as *vera causa* in observable domains, they may be extended to other domains of a common natural kind. In fact, as shown in the previous section, Darwin's inquiries from 1837 to 1839 actually enjoyed both kinds of evidence.

In sum, Achinstein's suggestion that arguments to explanations or to the best explanations have no epistemic force needs qualification. We might concede that in Darwin's case just the observations of Galapagos finches is alone insufficient to justify belief in his suppositions about the causal structure of local speciation, just as good explanations of everyday action are not given evidence when applied to one instance. However, Darwin also observed local species of mockingbird, armadillo, ostrich, llama, and others, indicating that his explanatory scheme was not only conceptually non-ad hoc, that it is addressed to more than just one case that might have initiated the search for an explanation, but also it is empirically non ad hoc as well, that is it is confirmed in idependently empirically certifiable circumstances. This independently empirically certified application of his explanatory scheme is as much evidence, as are the inductive arguments cited above, that biogeography is in respect of speciation lawlike in its structure. Actually the finch case alone would indicate that genetic evolutionary laws are not empirically ad hoc, for they explain differences in the beaks of several related species in different ecological circumstances, each of which is empirically certified independently. Analogously, if large cash awards are known empirically to be commonly awarded for writing books in philosophy, then that such an offer would explain my present efforts would be evidence for this hypothesis, which is an instantiation of a pattern already recognized to occur or to be likely. Also if other behaviors are explained by this hypothesis, e.g. my borrowing large sums at the time I write, then it will warrant the hypothesis, particularly if some of these other behaviors exclude some of the rival explanatory hypotheses.

Thus appealing to the explanatory power of an hypothesis has no evidential import if (i) it does not explain actual phenomena and (ii) there is no reason to believe that the domain of its application has a lawlike causal or ontic structure. Achinstein's view that the appeal to the explanatory power that a hypothesis would have, if true, is not a form of evidence seems correct. Also his suggestion that empirically ad hoc applications of explanatory hypotheses, i.e. the demonstration that H explains just one empirically certifiable phenomenon P, has no epistemic import seems plausible, for even though boundary conditions C can be observed independently of P, this one application of H does not demonstrate the lawlike character of the domain in which C and P occur. But his claim that other explanatory applications of H are not evidence seems just false, at least by the criteria for explanatory credibility developed above. If H is applied to circumstances and phenomena C′,P′, which are respectively independently empirically certifiable from C,P, then evidence is provided that the domain is lawlike. That is, there is a law in the domain which, when applied in different circumstances, gives different outcomes. Once it is established that a domain has a lawlike causal structure in respects relevant to the concepts used in H, that H explains P under C is a reason for believing H.

Even though explanatory sufficiency in one empirical application is insufficient to warrant belief, it does not follow that this sufficiency is not the kind of evidence that can, in other circumstances, support belief. In short, though in some circumstances certain kinds of evidence warrant only consideration and not acceptance, that evidence is still of a epistemic kind. This should be particularly plausible if it is recalled that in epistemic research we are looking for believable hypotheses, so any initial induction of H's believability is *prima facie* evidence that it is worth pursuing as a means to extending knowledge, for it suggests that such pursuits might be successful.

Turning to some of Achinstein's other criteria for consideration, one is that H answers questions generated by some program of research. But, as argued in previous chapters, programs of research generate certain questions, such as those demanding specification of forces and velocities in the molecules of a gas, because it is epistemically important to find testable explanatory applications of the concepts and laws at the core of the program. Though providing specifications of these mechanical variables does not itself establish the hypothetical values as worthy of belief, when the model thus developed is empirically confirmed it is more believable by virtue of its being a model of a recognized research program, kinetic-molecular theory, than if it were simply a realization of hypothetical mechanical values of some model invented for just these phenomena. Otherwise one is driven to endorse the credibility of *ad hoc* mechanical explanations, something which Newton, Maxwell and others sought to avoid (Kavaloski, 1974). Entertaining hypotheses to criticize them, to show their possibility, to construct approximative models with unrealistic assumptions, equally bring to bear epistemic considerations. Hypotheses worthy of criticism are *prima facie* believable and they may be criticized as a means to eliminating candidates for truth. Possibility is a necessary, though not a sufficient condition for credibility. Thus whatever conditions show a hypothesis possible, e.g. consistency or coherence with background belief, contribute to arguments for its truth. Approximative models may be chosen as means to the development of more realistic or empirically accurate models. In mechanics exact computations can be given only for two-body models and three, four, ... body models must be constructed by superimposing purturbations on two, three,... body models respectively. Also approximative models may be the means of isolating for discovery causes of observed perturbations, as happened in the discovery of Neptune. Approximative models thus often occur in epistemically directed heuristics, and the warrant of their use is their effectiveness in achieving epistemic ends. Thus Achinstein has not established that there are special heuristic considerations distinct from or not founded upon epistemic criteria.

The examples from heat theories and optical theories that Achinstein discusses (1987a,b) suggest that initial evidence for one explanatory scheme can be provided by eliminative arguments excluding alternative schemes. Darwin's arguments against creationism parallel Achinstein's account of Maxwell's initial evidence for the postulates of kinetic theory. Of the two initially credible explanatory schemes for speciation, special creation and reproduction with adaptation, the first is excluded because its mode of action is mysterious in the various actual circumstances contemplated for its application. How, Darwin asks, does the creative force's influence spread from its origins, the continents? Does it prevail like a halo around continents before islands are formed (B98)? We can observe the mode of action in natural means of colonization in the occasional arrival of accidental birds in remote places. Variation can be observed in domestic birds. Hence conditions for the application of natural laws are accessible and have been accessed, whereas these conditions for creative forces, the particular plan God might have for inhabitants of Charles Island in the Galapagos, are not thus credible.

Thus, as does Maxwell and others, Darwin provides some of his initial evidence for reproductive evolution by eliminative inductive reasoning. However Achinstein does not consider how the major premise in this kind of argument is established, e.g. that light is propagated either by motions of some material substance or by motions within an unmoving substance, that species are produced by creation or by reproduction with modification. For the eliminative argument to have epistemic import its premises must be worthy of belief. Achinstein's position would exclude abductive arguments to such premises, so other kinds of evidence must be sought in defense of his position.

In the commonsense explanations of action Achinstein cites against abduction, many good explanations are conceivable, but none has prior empirical evidence that would make it stand out as especially credible. In Darwin's and Maxwell's circumstances only two kinds of explanation are initially plausible, respectively speciation by creation or by reproduction with modification and heat as a substance or heat as motion. The initial plausibility cannot be based on evidence that these are the only two hypotheses worth consideration, for then the exclusion of one would give us only the considerability, not the believability, of the remaining hypothesis. Such initial plausibility could be by inability to think of further alternatives, which can hardly be considered positive evidence. This inability can be corrected by further conceptual development, which could well lead to concepts more fitting to the circumstances than those initially considered in the disjunctive premise. Also, contrary to the claims of one abductivist, Peirce, there is no reason to think that our conceptual abilities are adapted *a priori* just to the appreciation of all actual causal structures in the external environment. Any such adaptation can reasonably be expected in the realm of environmental interactions that influenced the natural selection of mental capacities, but causal patterns that occur in long term geological history, the domains of astronomy, nuclear and atomic physics can hardly be expected to be imprinted *a priori*. Thus *a priori* restriction to Newtonian causal architecture and the explanatory categories appropriate for purposive action among humans excludes, on psychologistic grounds, other possible causal architectures, and it should be hoped that human inquirers are capable of transcending these psychological barriers.

There remains the possibility that there is inductive evidence for the initial credibility of the disjunctions between modes of causal transmission through space, in the case of the light theories, or for some of the assumptions underlying Maxwell's kinetic theory. It has already been suggested that the initial evidence for evolution by adaptation in reproduction is inductive, that is it is drawn from observations indicating that variation occurs in reproduction among domestic species and that genetic relations among species can be composed of genetic relations among individuals. Thus an inductive bridge is provided from genetic relations causing similarities within a family of individuals or a clan to genetic relations causing similarities within a genus. In the case of theories of light the initial induction is from observation of processes by which action is transmitted in the observable world. Thus Huygens (1675) cites the propagation of sound and water waves as establishing that there is a mode of action in which propagation occurs without movement of materials composing the medium. This mode of propagation is offered as an alternative to the kind of action propagated by means of material projectiles. In the case of heat, there is a pattern of inference going back to Bacon that begins from the observation that heat is constantly associated only with motion. Increased motion produces and correlates with more heat. Since many heated bodies are not themselves in motion, the motion associated with heat must be attributed to bodies' invisible component parts. This inductive procedure resembles that contemplated by Newton in his Rule III, viz. if attributes are retained under operations of division, as are the mechanical attributes, then there are grounds for extending laws applying to the wholes to the parts. Accordingly, the invisible parts of bodies can have varying degrees of motion and can influence the invisible parts of other bodies by laws and forces observable in the whole bodies.

To summarize, many cases in the history of science suggest that initial evidence for explanatory hypotheses is inductive. The concepts or the laws implicit in the hypotheses have application to domains within the reach of experiment and observation and thus represent *vera causae*. However, this point does not exclude the possibility that hypotheses can, by virtue of their explanatory power, obtain initial evidence where such inductive evidence is not available. This source of evidence allows that hypotheses may be sought, worked on or believed whose terms and laws have no application in observable domains. Thus, for example, hypotheses in special non-mechanical terms, such as strangeness, color, and charm, and corresponding non-mechanical symmetry principles, may be considered and established in fundamental particle physics without inductive evidence.

6.6. CONCLUSION

In this Chapter traditional scientific methods were considered as heuristics, that is, procedures with applicability limited to certain kinds of problem in science. As Wimsatt pointed out (1980) each heuristic has a bias, that is, includes a substantive commitment concerning the subject of inquiry or peripherally related subjects. There is a cost-benefit tradeoff between the risks taken in embracing a biased heuristic and in obtaining sufficient information to conduct a program of inquiry with deliberate purpose and choice of the best instrumentalities for achieving that purpose. The bias renders it possible that the heuristic will mislead researchers, and, as Wimsatt claims, it follows that every heuristic will break down somewhere. Still a form of pragmatic rationality similar to Reichenbach's justification of induction suggests that biased heuristics are worth accepting because the alternative is a-heuristic trial and error at best and agnostic passivity at worst. Accepting the biases that cohere with what we take to be our best supported knowledge of a domain and its periphery and employing them in acts of further inquiry is the only way in which we can deliberately investigate our surroundings with the best rational hope of discovering epistemically and pragmatically significant items therein.

Among the traditional methods considered as heuristics are the method of analysis, various inductive methods, hypothetico-deductivism and abductivism.

The method of analysis has various components and interpretations. However the two most important components are the analysis of compounds into components and the regression from hypotheses in question to evidence. Both components are utilized in the traditional mathematical and logical applications of the method: The aim of analysis of proofs and geometrical figures into components is the hope of finding a component with known properties that can be used in the synthesis, the construction of a proof. These known properties can either themselves provide premises or can lead to premises either by knowledge at hand or by further steps in analysis. The regression is a strategy of inferring evidence from the form or content of the item to be proved. Analysis in this sense this can be conducted either directly, as in logic by determining applications of rules of inference from the form of the conclusion and using these to search for premises that would constitute the final step in the proof, and so on for intermediate steps. A less direct form of analysis identifies components in a geometric figure whose properties suggest evidence or intermediate steps in the proof that is sought. In either case analysis into components plays a role: In logic it is hoped that the regression will reveal component proof procedures that are already familiar. In geometry it is hoped that analysis of the figure will reveal components whose properties are either known or the proof of which is known or can be sought once required conclusions are found, where these known properties and proofs will also be components of the final proof once synthesized.

The strategy of regression can also be used effectively in natural science, as Descartes and Newton suggested. In this case analysis of compounds into components is again viewed as a means of generating evidence and is conducted in the hope of discovering either self evident properties of the components or properties that could be established by experiment and induction.

There are two different biases in this method, first that the conclusion is true and second that properties of the whole described in the conclusion can be derived from properties discovered in the parts. Pappus suggested that the first bias could be avoided by allowing the possibility that the analysis would lead to a falsity rather than a truth, in which case the conclusion could be refuted. Thus the search for evidence is not biased by constraining it to positive evidence for the conclusion. This strategy for demonstrating falsity works only if the steps in the regression are reversible, as they often are in geometry when one is inferring using equalities and similarities. However, it will not always generate refutations in logic wherein the steps in the proof sought are often entailments from premise to conclusion and not logical equivalences. Still, in logic one has at hand a number of rules of inference that can be used in search of premises and the non-tautologous premises thus generated can be posed as whether-questions (Hintikka and Bachman, 1991). Accordingly such investigations need not be biased in favor of the conclusion.

An extension of the method of analysis which also avoids this bias is to formulate alternative conclusions and then to search by analysis for evidence bearing upon each. In this variant the alternative conclusions may be constrained as possible or plausible answers to a favored question. However biases are still unavoidable: The list of answers considered is not always exhaustive, and when it is, the concepts giving sense to the alternatives considered is a bias. Second, many questions have substantive presuppositions that must be accepted as true for the question to be valid, and background knowledge structures and values are presupposed in assigning priority to a question.

A further bias in some variants of the method of analysis is its reductionism, or the belief that a whole can best be known in terms of properties of its components. The bias here is well known as the fallacy of composition: It is only under special circumstances that knowledge of parts contribute to knowledge of wholes. Recent studies of systems that interact non-linearly have exemplified the limitations of reductionism, though they have not excluded it as an often powerful heuristic.

Baconian inductivism has sometimes been viewed as a method of conducting an investigation without preconception. However, as Hume pointed out even primitive enumerative induction carries the bias that a component or aspect of nature is uniform or homogeneous. Furthermore the method of agreement must be described as a method of elimination, where the elimination is from a list of plausible causes. Otherwise the method is not heuristically practicable, but the contents of the list constitutes a bias. If the method of difference is described as reenforcing credibility of supposed causes that cannot be eliminated, it is subordinate to the method of agreement and accordingly shares its biases. The possibility always remains open that the remaining correlated qualities are effects of a common cause that has not been considered among those correlates tested. Inductive methods of covariation and curve-fitting are equally indecisive in their support of a conclusion and are heuristically efficacious only with a bias toward certain kinds of quantitative relations, such as linear, quadratic, exponential or other functions.

These inductive methods are also not considered heuristically efficacious in discovering fundamental causal mechanisms that underlie the correlations they produce. Newton, uniformitarian geologists, Darwin and many others have employed a procedure variously called the *vera causa* method and transduction. This method prescribes the extension of causal mechanisms established in observable domains to those in which only effects are observable and the effects are sufficiently similar to those in the observable domain that one is warranted in inferring similar causes. Hume (1980) has pointed out that this method requires strong similarities among the effects to provide assurances that unobservable causes are similar to those observed. Newton, Darwin and the uniformitarian geologists successfully applied the method in cases in which the effects compared were the same natural kinds, e.g. mechanical systems and sexually reproducing organisms. The assumption that effects in two domains are a common natural kind with all the baggage concerning what counts as a natural kind is a bias, though it is the kind of bias which provides the method with credible conclusions that might otherwise be doubtful. It has also been suggested that the method can be extended beyond common natural kinds to systems which, though characterized in distinct physical quantities, have like spatiotemporal structures.

Finally it was argued that the method of abduction was considered by Hanson to avoid the a-heuristic character of the hypothetico-deductive method by restricting hypotheses to those whose explanatory power could be acknowledged. Explanatory power is obtained when a law or internally coherent theory is so formulated that a variety of distinct events, entities or processes can be inferred from it under various initial and boundary conditions. Abduction based on explanatory power thus understood is biased with the assumption that the domain to which it is applied has the dynamic or ontic structure such that what exists or happens in the domain consists of various variations upon a common theme, as it were. This is a kind of 'uniformity in nature' that is more subtle than the more straightforward homogeneity presupposed by enumerative inductive methods, though it can be regarded as one of the various structural patterns of a domain of nature that have emerged in our reflections upon transduction. It appears that the power of transductive heuristics is the provision of grounds for assuming that a newly investigated domain has the causal or the ontic architecture that would allow us to presume that it is amenable to a certain kind of explanation. Once a hypothesis is formulated, abductive argumentation is available as grounds for credibility or at least the hope that the hypothesis, once further articulated and tested, will be credible.

CHAPTER SEVEN

SUMMARY AND CONCLUSIONS

1. The 'logic of discovery' may be considered a 'logic' because it contains normative rules that guide thought and research in science. However it does not consist entirely of universal or necessary rules, rules constitutive of science the violation of which would be 'unscientific'. In view of alternatives to classical logic and the reasonable suspension of any rule, whether of strict logic or of scientific method in the traditional sense, as well heuristic rules, there is no such thing as a constitutive rule for science. Most of the guiding principles for scientific research, whether conceptual or empirical, are local heuristics with contingent commitments or presuppositions. Most of the important metascientific rules and values cannot be derived a priori from deductive logic or mathematical probability theory, but rather require a posteriori justification. Such justification often comes by exemplification in research that is judged successful on other grounds, as can be found in the history of science.

2. Deterministic algorithmic rules whose application guarantees a product, that is, effective procedures, can, in certain circumstances produce laws, concepts and other items of scientific knowledge. However there is no reason to believe that these will solve all scientific problems because their application presupposes interpretation and scientific inference in gathering and arranging 'data' and thus far is limited to 'discovering' rather elementary functions. Algorithmic procedures are 'heuristics' in the sense that they constrain search spaces, their applicability is limited to certain kinds of epistemic condition and they contain metascientific biases. The objectives programmed into these search programs are drawn from cases in the history of science in which the goals have

already been achieved. Algorithms for generating metascientific objectives *de novo* do not seem plausible. Still scientific problems can be divided into subproblems some of which may be amenable to algorithmic heuristics, and thus the speed and power of computers can be of considerable assistance to the scientist, though it cannot be expected to replace her. Heuristics for the computer can be drawn from historical as well as psychological, logical and mathematical research.

3. A lexical definition of "discovery" is as follows:

> 1. To reveal by divulging, as a secret, by making known the identity of, as a person, or by laying open to view, as a thing hidden or covered; to expose; to bring to light...
>
> 2. To uncover, to remove or lift (any covering).
>
> ***
>
> 4. To obtain for the first time sight or knowledge of, as a thing existing already, but not perceived or known; to find out; to ascertain; also, to espy; detect; descry. (Merriam-Webster, 1936)

The 'uncovering' or 'laying open to view' is an important component of this definition. If interpreted as a metaphor, as is the intent here, these terms can be applied to science. Scientific concepts, theories and instruments enable us to comprehend, infer, manipulate, detect and image. Thus they open to our cognition various items otherwise inaccessible, and normative guidelines for doing so constitute at least part of a 'logic of discovery'. Secondly 'ascertaining' is a component, and as applied to science the process of discovering should also include procedures for justification. In both of these cases the lexical definition sanctions efforts to find evidential relations and to produce evidence satisfying these relations. Thirdly, 'sight or knowledge of something existing already' might suggest a contrast between 'discovery' and 'invention', where the latter term applies to the conception and production of something that does not already exist. However, Whewell and Kuhn both have argued convincingly that the invention that takes place in conceptualization is an important component of scientific discovery. Items that may already exist may not be investigated, perceived, uncovered or understood until investigators develop concepts, theories or purposes that give these items theoretical significance or practical importance.

All of these points cohere with the traditional concneption of knowledge as justified true belief. Conceptualization renders it possible to entertain a proposition, to consider it and its alternatives, and is necessary though not sufficient for considering possible evidence for it and its alternatives. 'Meanings' are produced when items become common knowledge in a community so that they may be packaged and linked to a terminology, a necessary condition for practicable communication and a product of the generation of scientific knowledge. The evidential relations sought in science are those causal and structural relations between credible or sensible items and those items in the domain of investigation. By means of these relations sensory perceptions and established beliefs will make it at least probable, if not conclusively support some belief about the latter items. Since discovery is new knowledge, the 'logic of discovery' should be concerned with conceptualization as well as the generation of justifying evidence. In sum, science is not properly divided into contexts of discovery and justification. Rather scientific research is conducted in the context of discovery, and presenting finished research reports designed rhetorically to persuade a community to accept, thereby generating meanings, is as much a part of this context as is initial conceptualization.

4. Kuhn is wrong in denying that scientists deliberately pursue novelty and affirming that serendipity is the only source of novelty. Examples from Darwin, Kepler, and molecular genetics demonstrate that sometimes scientists set specific goals of epistemic novelty and achieve something that fits those objectives, though not always exactly as anticipated. These cases do not ex-

clude serendipity in science, however.

Serendipitous discoveries occur when significant answers to a question Q' are produced as a consequence of pursuit of question Q. Circumstances pertaining to pursuit of Q are causally responsible for encounter with conditions giving rise to Q'. Thus by pursuing Q, Q' is raised and solved. Q is a means to solving Q', though only perceived to be so in retrospect, for it was not intentionally pursued with the solution to Q' in mind. Recognition of significance of Q' may require knowledge beyond what is required to accept and value Q, the kind of knowledge summarized in cosmological background for scientific research. Without background knowledge beyond the domain of one's specialty one may not raise questions concerning circumstances serendipitously encountered, and thus serendipitous discoveries would be impossible.

5. Scientific problems pose epistemic objectives. Epistemic objectives are based upon perceptions of epistemic deficiencies. Metascientific beliefs M are one basis for these perceptions: A body of scientific belief may fall short on internal coherence or explanatory power, as Darwin perceived in creationist natural history. Also two bodies of belief may appear mutually incoherent, e.g. theoretical models T may fail to cohere with ontological or cosmological background B or with presumed empirical knowledge E, in which case either member of the pair may fall under doubt. Thus quantum theory fails to cohere with deterministic cosmologies and possibly also realistic metascience. Darwin found an incoherence between species concepts that exclude transmutation and the geographical representation of related species in the southern hemisphere.

Rational inquiry, contrary to positions taken by earlier historicist philosophers, excludes the wholesale rejection or general doubt about the ontological, conceptual, nomological, theoretical, metascientific, and empirical components of the body of scientific belief. Rather it must be based upon specific doubts addressed to only one or at most a few of these several components at a time. Thus ontological inquiries can be made about some conceptual or theoretical development in science in the light of metascientific or metaphilosophical standards of coherence. Similar inquiries may seek to revise existing metascientific belief or valuation. Conceptual inquiries can be made in the effort to explain some empirical belief or law. Other cases of explanatory inquiry may seek theoretical models, viz. descriptions of components of a system, its boundary and initial conditions, without seeking novelty in law, concept, or methodological value. Empirical inquiries may appeal to background ontologies for possible relations between domains of inquiry and domains that are empirically accessible, or may appeal to theoretical models and background ontologies for substantive relations that will render some happening observable or for guidelines for practical methods or recipes for generating observed evidence. In general, scientific inquiry uses substantive scientific belief as a tool for generating more knowledge.

6. According to formulations of erotetic logic held by Aqvist and Hintikka, questions describe and demand the achievement of epistemic objectives. Valid or true questions, that is, those whose presuppositions are credible, may also be appraised as more or less important. The epistemic component in this formulation has two useful features: The intensional epistemic operator can be indexed to designate the context of inquiry in which a question is posed, that is the particular ontology, conceptual and nomological framework, metascience and empirical capabilities characterizing a community within which the question is posed. The epistemic operator also expresses the objectives of epistemic inquiry, the achievement of knowledge of answers to these questions. Since questions' presuppositions are expressed without epistemic operators, aspiration to knowledge can be attributed to a historical figure in her own epistemic context without endorsement from the context in which the attribution is made.

Generally questions describe their answers by listing or by attributing properties without divulging these answers. In the terminology of some writers, they place constraints on possible answers, and the process of pursuing a question can employ heuristics that loosen or tighten these constraints. Also the constraints placed on a question, such as the explanation of the introduction of species, can change radically as a scientific field evolves. Many examples of scientific revolutions, such as the Keplerian revolution in astronomy, the Darwinian revolution in natural history and biogeography, revolutions in atomic mechanics and astrophysics, and the revolution of genetics in the second half of the 20th century, include significantly radical changes in the constraints placed on why-questions. Equivalently, these revolutions can be described as more or less radically altering explanatory objectives in these several fields.

Searching for an answer to a question can generate information incompatible with the question. Hence answers can be corrective. Erotetic logic provides a scheme in which such conflicts can be represented, but it offers no help in deciding whether the next step in a research program is to reject the answer or to seek a revised question and consequent revisions in the background knowledge relevant to the question's presupposition. Corrective answers to a question Q originally posed can be considered serendipitous discoveries for they are answers to a different question Q′.

7. The following principles govern question choice:

a. Questions can be falsified by rejecting their presuppositions. Comparison with falsificationism shows that this is insufficient as a methodology of question appraisal, particularly because there are in science decisions that require credibility or reliability judgment. Problems compatible with our beliefs need to be ordered as to preference, particularly with limited resources for research.

b. Deontic-epistemic logic shows that if knowledge of conditional is obligatory, then whatever priority attaches to its antecedent also attaches to its consequent.

c. The semantics of questions in the H-A formulation implies that principles for selecting questions are heuristic principles under the usual construal of 'heuristic': Given some actual knowledge KP, epistemic alternative worlds u contain KP and additional knowledge Kq. Such worlds can be countably or continuously infinite in number. Questions select among these alternative worlds some u+ and demand their pursuit. One cannot practicably pursue all of the infinite possibilities containing the various items of knowledge Kq. However a selection from these of items of knowledge to be sought is a narrowing of the possibilities to be investigated, and thus principles prescribing such selection fits the conception of heuristic commonly employed in artificial intelligence and elsewhere. An example of such a heuristic is Lakatos' positive heuristic combined with the method of analysis. This heuristic prescribes knowledge of individuals that instantiate a known law a means to obtaining specific theoretical models of these individuals. Thus one might demand knowledge of one or several individual planets as means to producing a Newtonian model of their trajectories.

d. With the aid of Glymour's and Bayes' theories of evidence and certain other principles, such as the principle of unbiased evidential inquiry and consilience, it can be shown that the importance of knowing a hypothesis transfers to the importance of knowing evidence for that hypothesis. Thus the importance of the question "Is h true?" transfers without diminution to "Is e true?" where e is evidence for or against hypothesis h.

Baysean conditions for evidence place knowledge of several terms at a high level of import, particularly the inverse term $P(e_j,h_k\&e_i)$ for favorable and unfavorable e_j, and independent evidence $P(e_i,e_j)$ or $P(e_j,h_m\&e_i)$ for rival hypotheses h_m. These amount to valuing the knowledge of certain evidential relations, in this case natural probabilities. The search for this knowledge is warranted by the Principle of Unbiased Evidential Inquiry, but is also supported by the import of consilience and by Glymour's requirement for evidence. These principles also permit the inference from valuations of questions $!(K\langle i\rangle h \vee K\langle i\rangle{\sim}h)$ to $!(K\langle i\rangle e \vee K\langle i\rangle{\sim}e)$ for hypothesis h and evidence e.

e. Problems that bear upon the credibility of archetypal beliefs, e.g. the belief that species are stable, that gravity is inverse square, that physical processes are deterministic, that biological processes are physio-chemical, are particularly important.

f. Dialectics play a role in generating problems, particularly problems raised by rivalries between research programs.

Epistemically offensive strategies offer new ways of solving problems P′ unsolved by received programs and strive to solve problems P″ solved by the old programs but unsolved by the upstart programs. Epistemically defensive strategies seek to solve or dissolve problems P′ solved by upstart programs.

In the dialectic process some problems, empirical or otherwise, can become crucial in deciding between rival theories. They are thus highlighted as of outstanding importance.

g. Problem generality, promise of solving a number of problems at once can highlight certain formulation or approach to a problem.

h. Time can erode confidence of power of received paradigm or research program, particularly if a problem important on other grounds is left unsolved.

i. Possession of conceptual or technological tools for addressing a problem can raise the importance of that problem. However, contrary to claims of many, lack of available means does not necessarily mean a problem must be put on back burner. Under appropriate circumstances resources can be invested in search for tools for solution.

8. The interfield approach to relevance of computational replication of scientific discoveries is this: The computer is an aid to inquiry, but present day computers are not adequate to represent all aspects of human thought. The search for heuristics appropriately draws on psychological and historical studies as well as computational efficacy. Heuristics may be appraised inductively in reference to their track record, by analogy to heuristics of known efficacy, or by means of judgments of efficacy from the heuristics' intrinsic characteristics.

Simon's heuristics are general in the sense that they are subject-matter independent. However they do not form a 'logic' because their applicability is limited to certain kinds of initial condition and they can effectively solve only certain problems. They do demonstrate that the introduction of new concepts, in this case 'intrinsic properties' required for the formulation of laws, can be done by a programmable routine under limited circumstances. However, these routines are discovered and justified a posteriori by their success in other domains of inquiry.

Contrary to claims by Simon *et. al.*, the BACON programs are not actually data-driven or inductivistic. Rather they are driven by the search for laws, that is functions of the data that are constant in value. The nature and form of a law is derived from historical exemplars, e.g. laws of uniform rotation in astronomy. Hence the success of these programs in no way supports inductivism, the view that scientific laws and theories can be generated in a routine way from appropriately collected empirical 'data'.

9. PROCESS OF INDUCTION (PI) programs operate on the assumption that induction is based upon prior substantive belief regarding the uniformity of certain domains or categories. The strategy for problem solving is one of applying in succession items of knowledge that are graded as to their relevance to a problem. Relevance relations are subject to revision as the system gains experience in problem solving. Initial failures are followed by activation of successively less relevant items of belief. Concepts are nodes in ordered (default) hierarchies of structural, causal and pragmatic relations and can be formed by combining other concepts and dropping less essential relations in the case of conflict. PI is open to parallel research strategies. Though human individuals are not capable of consciously carrying out very many tasks in parallel, communities are, though the heuristic advantages gained by such hedging comes at a price in the number of resources utilized. Parallel strategies are exemplified in many contexts and illustrate analogous heuristics that operate upon a common principle, in this case the notion that probability of failure can be multiplied for parallel systems. Generalization and specialization are further conceptual heuristics. One form of generalization is the merging of concepts: Varieties and species are merged together by removing distinct accounts of their origins. Concepts that fail can be specialized by adding qualifiers to their conditions: Birds can fly, but penguins cannot. Hence the rule 'If X has feathers, X can fly' is replaced by 'If X lacks penguin characters but has feathers, X can fly'.

10. The following are metaheuristic principles:
a. Analogies between an unknown and a successful heuristic, viz. with those generators that have produced empirically and explanatorily successful theories, can support the use of the former heuris-

tic without knowledge of how it actually works.
b. The argument from analogy can be strengthened by discovering the basis for the analogies. The success of a heuristic is due to heuristically significant properties, viz. properties whereby in a set of circumstances a heuristic will constrain the problem space for practicable search. Knowledge of these properties alone is sufficient to justify a heuristic in appropriate circumstances without knowledge of similar heuristics.

A heuristic is justified if its use is warranted by a balance of costs and benefits in a given set of circumstances. However warranted use is no guarantee that the heuristic will be successful in this instance.

11. Feyerabend's attack on 'method' can be interpreted as supporting the conclusion that there are no constitutive rules for science. Instead the 'rules' of scientific method are heuristics. In some circumstances they can usefully constrain the options facing a researcher, though in other circumstances they can obstruct or mislead creative research. Thus without suitable evidential relations, the Popperian demand for falsifiability might have to be suspended, thus allowing one to entertain for further research a hypothesis that cannot at present be tested. However this Popperian rule also sanctions the importance of a problem raised by such an hypothesis, a problem of determining what evidence can be relevant to accepting or rejecting the hypothesis. Similarly the non-ad hocness rule need not exclude a hypothesis but rather can sanction the search for further theoretical models employing the hypothesis or for further independent empirical tests of it.

The method of analysis described by Pappus prescribes examining the form or content of a conclusion whose truth is in question and then, in the light of this examination, asking what could prove it or could provide evidence for it. This method can be construed as a powerful heuristic for generating questions in an investigation: From the question "Is h true?" can be generated "What kind of e would establish or refute h", "How is e made evident to cognition or the senses?" and "Is e true?".

Breaking a whole down into components, as in finding components of a geometrical figure whose properties are known, or in analyzing light into component colors, is a subheuristic to the regression to evidence procedure. The strategy is based on the assumptions that the whole can be known in terms of its components and the components are already known or are accessible to knowledge. These are contingent assumptions, and where they break down analysis cannot be used in the search for evidence. Reductionism makes just these assumptions and thus it can be considered a heuristic for obtaining knowledge (Wimsatt, 1980). Another utilization of this procedure of division into components is the separation of essential causes from peripheral contingencies, as implemented in the succession of Newtonian models for the Solar System beginning with a two body system consisting of the Sun and a planet and then adding other planets with their purturbations of this first system. Here again the component two-body system is more readily known than the whole system of six or nine planets with the secondaries. The 'synthesis' side of the procedure is the production of a succession of more realistic models with successively more influential purturbations included.

12. There are a variety of inductive methods beyond simple enumerative induction, including Mill's methods of agreement, difference, concomitant variation and residue. In these credibility can rest upon elimination of plausible alternatives, uneliminability, covariation and exhaustive causal relations as well as upon numbers of instances. These methods are biased in the list of plausible alternatives that are eliminated. They will break down where actual causes are not included in this list. Inductivism is the view that such methods are constitutive rules for science. However, if they are considered as heuristics their utility in science can be acknowledged without embracing the anti-theoretical stance of traditional inductivism.

Newton's inductivism is different in its implications from that of Mill and certain positivists. It is a form of inductivism that does not restrict science to phenomenalistic theories. His method sets certain conditions for extending terminology and laws from observed to partially observable and unobservable domains such as celestial and atomic domains. Its credibility rests upon presumed causal homogeneity of space of both cosmic and atomic dimensions. Terms can be extended from whole to part, the heuristic of division, or from part to whole, the heuristic of composition. The heuristics of division and composition can serve as subheuristics to the methods of analysis and synthesis, for they may be considered fallible means of knowing part from whole or whole from part. Darwin's reasoning regarding the conditions of survival of populations, particularly regarding causes of death and failure of reproduction, makes use of the composition heuristic. Causes and other concepts applicable to individuals are applied to populations. Uniformitarian geologists such as Hutton, Lyell and Darwin extend Newton's inductive method to the temporal dimension presuming the causal homogeneity of geological time. However, this vera causa method can and has broken down insofar geological history extends into times in which physical conditions were radically different from those at present.

13. Hanson, Salmon and others distinguish the logic of discovery as concerning reasons specifically appropriate for entertaining but not believing a hypothesis. Although we can acknowledge Hanson's point that there is a difference between entertaining and believing a hypothesis, there are no specific kinds of reason appropriate for either. Rather, initial or weak criteria for belief can provide grounds for entertaining a hypothesis, whether they be inductive or abductive. The hypotheses worth pursuing are those that we have reason to think might turn out to be true and credible, and these reasons, in other circumstances, can be reasons for believing. Achinstein's argument that explanatory power cannot be construed as evidence for truth would seem to undermine this point if explanatory power can be considered to be grounds for entertaining. However, explanatory power must be demonstrated by application to different instances, and such application can be considered consilient evidence if the instances are actual. Hence demonstrated explanatory power can be evidence for truth.

14. Herschel's method of analogy can be applied where the *vera*

causa method breaks down, that is where descriptive terms cannot be extended from one to another domain. Promising analogies can be constrained by requiring that the domains compared have like attributes that occur in clusters thereby allowing them to be considered members of a common natural kind. These are 'real analogies'. Domains of distinct natural kinds might still fruitfully be considered physically analogous if the spatiotemporal characteristics of causal propagation are alike, as in field theories, even though in detail the forces in the two domains are distinct.

15. The hypothetico-deductive method (HD) can also be considered a heuristic, though many of its advocates were hostile to the possibility of a logic of discovery. In contrast to constructive methods for conceptualization advocated by positivists, HD permits maximum flexibility in introducing novel concepts. However HD is a-heuristic in that it offers no guidelines or constraints upon novel concepts beyond the prescription that methods for testing such concepts be sought. Hanson's account of abduction adds a restriction to hypotheses to be pursued: Their explanatory power should be initially manifest. One heuristic weakness of this proposal is that some conceptual development is required for explanatory power to be evident, and Hanson's heuristic provides no means of deciding initially which concept is worthy of such development.

16. Traditional methodologies considered as heuristics have substantive biases, that is, contingent commitments about the subject of inquiry or peripherally related domains. Inductive heuristics presuppose a homogeneous domain or certain specific uniformities between domains, the latter of which can be in various terms. For example a domain might be homogeneous in terms of spatiotemporal structure, dynamic causal relations or more abstract properties such as energy conservation and entropy increase. Abductive heuristics presuppose that some attributes in a domain will be lawlike in their occurrence, and as in the case of homogeneities these attributes can be various in various domains. Heuristics of composition and division presuppose homogeneities between com-

pound and component domains. The greater this commitment the greater its heuristic value, for more information about a domain provides greater constraints upon possible outcomes of inquiry. Risk of error thus rises with heuristic value. Nevertheless it can be argued that undertaking such risks is rational on grounds that the gains in the practicability of a research program outweigh the risk of error and failure. Generally risking error by assuming the bias of some heuristic is the price of deliberate action in the search for knowledge.

REFERENCES

Achinstein, P.: 1987, 'Light Hypotheses', *Studies in History and Philosophy of Science* **18**:293–337

Alvarez, W. and Asaro, F.: 1990, 'An Extraterrestrial Impact', *Scientific American* **263**:78–84

Aqvist, L.: 1975, *A New Approach to the Logical Theory of Interrogatives*. Tubingen: TBL Verlag Gunter Narr

Aristotle: 1941, 'Posterior Analytics', in McKeon, R. (ed.) *The Basic Works of Aristotle*. New York, NY: Random House

Bechtel, W.: 1986, 'What Happens to Accounts of Mind-Brain Relations if We Forgo an Architecture of Rules and Representations', *PSA 1986*, Fine, A. and Machamer, P. (eds.) East Lansing, MI: Philosophy of Science Association

Bechtel, W.: 1988, *Philosophy of Science: An Overview for Cognitive Science*. Hillsdale, NJ: Lawrence Erlbaum Associates

Beckner, M.: 1959, *The Biological Way of Thought*. Berkeley: University of California Press

Belknap, N. and Steel, T.: 1976, *The Logic of Questions and Answers*. New Haven, CT: Yale University Press

Blanchowicz, J.: 1989, 'Discovery and Ampliative Inference', *Philosophy of Science* **56**:438–62

Bromberger, S.: 1965, 'An Approach to Explanation', in Butler, R.J. (ed.) *Analytical Philosophy*. Second Series. Oxford: Basil Blackwell

Bromberger, S.: 1966, 'Why-Questions', in R. Colodny (ed.) *Mind and Cosmos*. Pittsburgh, PA: University of Pittsburgh Press

Bromberger, S.: 1971, 'Science and the Forms of Ignorance', in Nagel, E., Bromberger, S. and Grunbaum, A. *Observation and Theory In Science*. Baltimore: Johns Hopkins Press

Bromberger, S.: 1988, 'Rational Ignorance', *Synthese* **74**:47–64

Browne, J.: 1983, *The Secular Ark: Studies in the History of Biogeography*. New Haven: Yale University Press

Brush, S.: 1989, 'Prediction and Theory Evaluation: The Case of Light Bending', *Science* **246**:1124–1139

Buchanan, B.: 1985, 'Steps Toward Mechanizing Discovery', in Schaffner, K. (ed.) [1985]

Buchanan, B.: 1982, 'Mechanizing the Search for Explanatory Hypotheses',

Asquith, P. and Nickles, T. (eds.) *PSA 1982*. East Lansing, MI: Philosophy of Science Association

Bunge, M.: 1967, *Scientific Research* Vols. I & II. Berlin: Springer-Verlag

Campbell, D.T. (1974a) 'Unjustified Variation and Selective Retention in Scientific Discovery', in Ayala, F. and Dobzhansky, T. (eds.) *Studies in the Philosophy of Biology*. London: Macmillan, 139–161

Campbell, D.T. (1974b) 'Evolutionary Epistemology', in Schilpp, P.A. (ed.) *The Philosophy of Karl Popper*. LaSalle: Open Court

Capek, M.: 1961, *The Philosophical Impact of Contemporary Physics*. Princeton, NJ: Van Nostrand

Carnap, R.: 1928, *Der Logische Aufbau der Welt*. Berlin: Welkreis-Verlag

Carnap, R.: 1950, *The Logical Foundations of Probability*. Chicago: University of Chicago Press

Chamberlin, T.C. (1981) 'On Multiple Hypotheses', in Tweeny, R.D. *et al.* (eds.) *On Scientific Thinking*. New York: Columbia University Press

Cohen, I.B. and Drabkin, A.E.: 1952, *A Source Book in Greek Science*. Cambridge, MA: Harvard University Press

Cohen, I.B.: 1980, *The Newtonian Revolution*. Cambridge: Cambridge University Press

Courtillot, V.E.: 1990, 'A Volcanic Eruption', *Scientific American* **263**:85–92

Cushing, J.: 1982, 'Models and Methodologies in Current, Theoretical High-Energy Physics', *Synthese* **50**:5–101

Darden, L.: 1980, 'Theory Construction in Genetics', Nickles, T. (ed.) *Scientific Discovery: Case Studies*. Dordrecht: D. Reidel, 151–170

Darden, L.: 1982, 'Artificial Intelligence and Philosophy of Science: Reasoning by Analogy in Theory Construction', Asquith, P. and Nickles, T. (eds.) *PSA 1982*, Vol. 2, East Lansing MI: Philosophy of Science Association, 147–165

Darden, L. and Cain, J.A.: 1989, 'Selection Type Theories', *Philosophy of Science* **56**:106–129

DeBeer, G.: 1960, (ed.) 'Darwin's Notebooks on Transmutation of Species', *Bulletin of the British Museum (Natural History)*, Historical Series **2**

Dewey, John.: 1938, *Logic, the Theory of Inquiry*. New York: Holt

Dewey, John.: 1939, *Theory of Valuation*. Chicago: University of Chicago Press

Einstein, A.: 1954, *Ideas and Opinions*. New York: Crown Publishing Co.

Feigenbaum, E.A., Buchanan, B. and Lederberg, J.: 1971, 'On Generality and Problem Solving: A Case Study Using the dendral Program', *Machine Intelligence* **7**:165–190

Feigenbaum, E.A. and McCorduck, P.: 1983, *The Fifth Generation*. Reading, MA: Addison-Wesley

Feyerabend, P.K.: 1962, 'Explanation, Reduction and Empiricism', in Feigl, H. and Maxwell, G. *Minnesota Studies in the Philosophy of Science*, Vol.

III. Minneapolis: University of Minnesota Press
Feyerabend, P.K.: 1974, *Against Method*. Atlantic Highlands: Humanities Press
Feyerabend, P.K.: 1987, *Farewell to Reason*. London: Verso
Galison, P.: 1987, *How Experiments End*. Chicago: University of Chicago Press
Giere, R.: 1988, *Explaining Science*. Chicago: University of Chicago Press
Glymour, C.: 1980, *Theory and Evidence*. Princeton, NJ: Princeton University Press
Glymour, C., Scheines, R., Spirtes, P. and Kelly, K.: 1987, *Discovering Causal Structure*. Orlando FL: Academic Press
Goldman, A.I.: 1978, 'Epistemics: The Regulative Theory of Cognition', *Journal of Philosophy*, **75**:509–523
Goldman, A.I.: 1986, *Epistemology and Cognition*. Cambridge MA: Harvard University Press
Goodman, N.: 1950, *The Structure of Appearance*. Cambridge, MA: Harvard University Press
Gruber, H. and Barrett, P.: 1974 *Darwin on Man*. New York, NY: Dutton
Hadamard, J.: 1909, *The Psychology of Invention in the Mathematical Field*. Princeton NJ: Princeton University Press
Hanson, N.R.: 1961a, *Patterns of Discovery*. Cambridge: Cambridge University Press
Hanson, N.R.: 1961b, 'Is there a Logic of Scientific Discovery?', in Feigl, H. and Maxwell, G. (eds.) *Current Issues in the Philosophy of Science*. New York: Holt Reinhart and Winston
Harre, R,: 1970, *The Principles of Scientific Thinking*. Chicago, IL: University of Chicago Press
Harre, R.: 1972, *The Philosophies of Science*. Oxford: Oxford University Press
Hempel, C.: 1954, 'A Logical Appraisal of Operationalism', *Scientific Monthly* 1:215–220
Hempel, C.: 1966, *Philosophy of Natural Science*. Englewood Cliffs, NJ: Prentice-Hall
Hesse, M.B.: 1962, *Forces and Fields*. New York, NY: Philosophical Library
Hesse, M.B.: 1966, *Models and Analogies in Science*. South Bend, IN: University of Notre Dame Press
Hesse, M.B.: 1974, *The Structure of Scientific Inference*. New York, NY: Macmillan
Hintikka, J. and Remes, U.: 1974, *The Method of Analysis: Its Geometrical Origin and Its General Significance*. Dordrecht: D. Reidel
Hintikka, J.: 1974, 'Practical vs. Theoretical Reason: An Ambiguous Legacy', in Hintikka, J. (ed.) *Knowledge and the Known: Historical Perspectives in Epistemology*. Dordrecht: D. Reidel, 80–97
Hintikka, J.: 1976, 'The Semantics of Questions and the Question of Seman-

tics', *Acta Philosophica Fennica* **28**:12–200

Hintikka, J.: 1978, 'A Discourse on Descartes' Method', in Hooker, M. (ed.) *Descartes: Critical and Interpretive Essays*. Baltimore: Johns Hopkins University Press

Hintikka, J.: 1981, 'On the Logic of an Interrogative Model of Scientific Inquiry', *Synthese* **47**:69–81

Hintikka, J. and Bachman, J.: 1991, *What if...? Toward Excellence in Reasoning*. Mountain View, CA: Mayfield Publishing Company

Herschel, J.F.W.: 1831, *Preliminary Discourse on the Study of Natural Philosophy*. London: Longman, Rees, Orme, Brown and Green

Holland, J., Holyoak, K., Nisbett, R. and Thagard, P.: 1986,, *Induction: Process of Inference, Learning and Discovery*. Cambridge, MA: MIT Press

Hull, D.: 1973, *Darwin and His Critics*. Cambridge, MA: Harvard University Press

Hull, D.: 1988, *Science as Process*. Chicago: University of Chicago Press

Hull, D.: 1989, *The Metaphysics of Evolution*. Albany, NY: SUNY Press

Hume, D.: 1980, *Dialogues Concerning Natural Religion*. Indianapolis, IN: Hackett Publishing Co.

Huygens, C.: 1675, *Treatise on Light*. Chicago: University of Chicago Press

Jardine, N.: 1976, 'Galileo's Road to Truth and the Demonstrative Regress', *Studies in History and Philosophy of Science* **7**:277–318

Jason, G.: 1988, *The Logic of Scientific Discovery*. New York: Peter Lang

Judson, H. F. *The Eighth Day of Creation*. New York: Simon and Schuster

Kantorovitch, A. and Ne'eman, Y. (1989) "Serendipity as a Source of Evolutionary Progress in Science", *Studies in History and Philosophy of Science* 20, pp. 505–530

Kavoloski, V.: 1974, 'The "vera causa" principle: An Historico-philosophical study of a metatheoretical concept from Newton through Darwin', Ph.D. diss. University of Chicago

Kelly, K.: 1987, 'The Logic of Discovery.' *Philosophy of Science* **54**:435–52

Keller, E.F.: 1990, 'Physics and the Emergence of Molecular Biology: A History of Cognitive and Political Strategy', *Journal of the History of Biology*, **23**:389–409

Kitcher, P.: 1983, *The Nature of Mathematical Knowledge*. New York: Oxford University Press

Kitcher, P.: 1985, 'Darwin's Achievement', in Rescher, N. (ed.) *Reason and Rationality in Natural Science*. Washington, DC: University Press of America

Kitcher, P. and Culp, S.: 1990, 'Theory Structure and Theory Change in Contemporary Molecular Biology', *British Journal for the Philosophy of Science*, **40**:459–484

Kleiner, S.A.: 1970, 'Erotetic Logic and the Structure of Scientific Revolution', *British Journal for the Philosophy of Science* **20**:149–165

Kleiner, S.A.: 1981, 'Problem Solving and Discovery in the Growth of Darwin's Theories of Evolution', *Synthese* **62**:119–162

Kleiner, S.A.: 1983a, 'An Aspect of the Logic of Discovery', *Southern Journal of Philosophy* **21**:513–536

Kleiner, S.A.: 1983b, 'A New Look at Kepler and Abductive Argument', *Studies in History and Philosophy of Science* **14**:279–313

Kleiner, S.A.: 1985a, 'Interrogatives, Problems and Scientific Inquiry', *Synthese* **62**:365–428

Kleiner, S.A.: 1985b, 'Darwin's and Wallace's Revolutionary Research Programme', *British Journal for the Philosophy of Science* **36**:367–392

Kleiner, S.A.: 1988, 'Erotetic Logic and Scientific Inquiry', *Synthese* **74**:19–46

Kleiner, S.A.: 1988, 'The Logic of Discovery and Darwin's Pre-Malthusian Researches', *Biology and Philosophy* **3**:293–315

Koertge, N.: 1982, 'Explaining Scientific Discovery', *PSA 1982*, Vol. 1. East Lansing, MI: Philosophy of Science Association, 14–28

Koestler, A.: 1959, *The Sleepwalkers*. New York: MacMillan

Koj, L.: 1989a, 'Inquiry into the Structure of Questions', in Koj and Wisniewski [1989]

Koj, L.: 1989b, 'On Justification of Questions', in Koj and Wisniewski [1989]

Koj, L.: 1989c, 'The Problem of Primitive Terms of the Logic of Questions', in Koj and Wisniewski [1989]

Koj, L. and Wisniewski, A.: 1989, *Inquiries into the Generating and Proper Use of Questions*. Lublin, Poland: Realizm Ragjonalnosc Relatywizm

Kosso, P. (1989) *Observability and Observation in Physical Science*. Dordrecht: Kluwer Academic Publishers

Kuhn, T.S.: 1959, *The Copernican Revolution*. New York: Random House

Kuhn, T.S.: 1970, *The Structure of Scientific Revolutions*. Chicago: University of Chicago Press

Kuhn, T.S.: 1977, *The Essential Tension*. Chicago: University of Chicago Press

Lakatos, I.: 1970, 'Falsification and the Methodology of Scientific Research Programmes', in Lakatos, I. and Musgrave, A. (eds.) *Criticism and the Growth of Knowledge*. Cambridge: Cambridge University Press

Lakatos, I.: 1978, 'Newton's Effect on Scientific Standards', in Worrall, J. and Currie, G. (eds.) *Imre Lakatos: Philosophical Papers*, Vol. 1. Cambridge: Cambridge University Press

Langley, P., Simon, H.A., Bradshaw, G. and Zytkow, J.: 1987, *Scientific Discovery*. Cambridge, Mass.: MIT Press

Laudan, L.: 1977, *Progress and Its Problems*. Berkeley: University of California Press

Laudan, L.: 1981, *Science and Hypothesis*. Dordrecht: D. Reidel

Laudan, L.: 1984, *Science and Values*. Berkeley: University of California Press

Lenat, H.: 1982, 'The Nature of Heuristics', *Artificial Intelligence* **19**:189–249

Lyell, C. (1831–3) *Principles of Geology*. London: John Murray
Mach, E.: 1919, *The Science of Mechanics*. Trans. Thomas J. McCormack, 4th ed. Chicago: Open Court
Maull, N.: 1977, 'Unifying Science Without Reduction', *Studies in History and Philosophy of Science* **8**:143–162
Maull, N. and Darden, L.: 1977, 'Interfield Theories', *Philosophy of Science* **44**:43–64
Maxwell, J.C.: 1876, 'General Considerations', in *The Scientific Papers of James Clerk Maxwell*. W.D. Niven (ed.) New York: Dover Publications, 1965
Maxwell, N.: 1974, 'The Rationality of Scientific Discovery', *Philosophy of Science* **41**:123–153, 247–295
McCauley, R.N. 'Intertheoretic Relations and the Future of Psychology', *Philosophy of Science* **53**:179–199
McGuire, J.E.: 1970, 'Atoms and the "Analogy of Nature": Newton's Third Rule of Philosophizing'. *Studies in History and Philosophy of Science* **1**:3–58
McKeon, R.: 1966, 'Philosophy and the Development of Scientific Methods'. *Journal of the History of Ideas* **27**:3–22
McKinnon, E.M.: 1982, *Scientific Explanation and Atomic Physics*. Chicago: University of Chicago Press
Mill, J.S.: 1893, *A System of Logic*. London: Longman's Green & Co.
Newell, A. and Simon, H.: 1972, *Human Problem Solving*. Englewood Cliffs, NJ: Prentice-Hall
Newton, I.: 1962, *Principia*. Motte, A. (trans.) revised by Cajori, F. Berkeley, CA: University of California Press
Newton, I.: 1730, *Opticks*, Fourth Edition. New York: Dover, 1952
Nickles, T.: 1980a, 'Scientific Discovery and the Future of Philosophy of Science', in Nickles, T. (ed.) *Scientific Discovery, Logic and Rationality*. Dordrecht: D. Reidel, 1–59
Nickles, T.: 1980b, 'Can Scientific Constraints be Violated Rationally?', in Nickles, T. (ed.) *Scientific Discovery, Logic and Rationality*. Dordrecht: D. Reidel, 285–315
Nickles, T.: 1985, 'Beyond Divorce: Current Status of the Discovery Debate', *Philosophy of Science* **52**:177–206
Nickles, T.: 1990, 'Discovery Logics', *Philosophica* **45**:7–32
Polanyi, M.: 1958, *Personal Knowledge*. Chicago: University of Chicago Press
Provine, W.: 1971, *The Origins of Theoretical Population Genetics*. Chicago: University of Chicago Press
Olby, R.C.: 1966, *The Origins of Mendelism*. London: Constable
Olby, R.C.: 1974, *The Path to the Double Helix*. Seattle: University of Washington Press
Oldroyd, D.: 1986, *The Arch of Knowledge*. New York: Methuen

Peirce, C.: 1957, *Essays in the Philosophy of Science*. Tomas, V. (ed.) New York, NY: Liberal Arts Press

Peirce, C.: 1960, *The Collected Works of C.S. Peirce*. Hartshorne, C. (ed.) Cambridge, MA: Harvard University Press

Pinch, T.J.: 1985, 'Towards an Analysis of Scientific Observation: The Externality and Evidential Significance of Observation Reports in Physics', *Social Studies of Science* **15**:1

Pinch, T.J.: Forthcoming, *Confronting Nature*. Dordrecht: D. Reidel

Popper, K.R.: 1961, *The Logic of Scientific Discovery*. New York: Science Editions

Popper, K.R.: 1983, *Realism and the Aim of Science*. Totowa, NJ: Rowman and Littlefield

Quine, W.: 1953, 'Two Dogmas of Empiricism', in Quine, W., *From a Logical Point of View*. New York: Harper Torchbooks

Reichenbach, H.: 1938, *Experience and Prediction*. Chicago: University of Chicago Press

Rescher, N.: 1984, *The Limits of Science*. Berkeley, CA: University of California Press

Rosenkrantz, R.: 1977, *Inference, Method and Decision*. Dordrecht: D. Reidel

Rudwick, M.: 1969, 'The Strategy of Lyell's Principles of Geology', *Isis* **61**:5–33

Ruse, M.: 1975a, 'Darwin's Debt to Philosophy', *Studies in the History and Philosophy of Science* **66**:159–81

Ruse, M.: 1975b, 'Charles Darwin and artificial selection', *J. Hist. of Ideas* **36**:339–50

Ruse, M.: 1977, 'Is Biology Different from Physics?', in Colodny, R. (ed.) *Logic, Laws and Life*. Pittsburgh, PA: University of Pittsburgh Press

Rummelhart, D.E., McClelland, J.R., et. al.: 1986, *Parallel Distributed Processing: Explorations in the Microstructure of Cognition*. Cambridge, MA: MIT Press

Roll-Hansen, N.: 1978, 'Drosophila Genetics: A Reductionist Research Program', *Journal of the History of Biology*. **11**:159–210

Salmon, W.: 1966, *The Foundations of Scientific Inference*. Pittsburgh, PA: University of Pittsburgh Press

Shapere, D.: 1966, 'Meaning and Scientific Change'. in Colodny, R. (ed.) *Mind and Cosmos*. Pittsburgh: University of Pittsburgh Press

Schaffner, K.: 1969, 'Correspondence Rules', *Philosophy of Science* **36**:280–290

Schaffner, K.: 1980, 'Theory Structure in the Biomedical Sciences' *Journal of Medicine and Philosophy* **5**:331–71

Schaffner, K.: 1985, (ed.) *Logic of Discovery and Diagnosis in Medicine*. Berkeley, CA: University of California Press

Simon, H.A.: 1977, *Models of Discovery*. Dordrecht: D. Reidel Publishing Co.

Shapiro, G.: 1986, *A Skeleton in the Darkroom*. San Francisco: Harper & Row

Stitch, S. and Nisbett, R.E.: 1980, 'Justification and the Psychology of Human Reasoning', *Philosophy of Science* **47**:188–202

Sulloway, F.: 1982a, 'Darwin and His Finches: The Evolution of a Legend', *Journal of the History of Biology* **15**:1–53

Sulloway, F.: 1982b, 'Darwin's Conversion: The Beagle Voyage and It's Aftermath', *Journal of the History of Biology* **15**:325–396

Thagard, P.: 1978, 'The Best Explanation and Criteria for Theory Choice', *Journal of Philosophy* **75**:76–92

Thagard, P. and Nisbitt, R.E.: 1983, 'Rationality and Charity', *Philosophy of Science* **50**:250–267

Thagard, P.: 1988, *Computational Philosophy of Science*. Cambridge, MA: Bradford Books

Thayer H.S. and Randall, J.H. (eds.): 1953, *Newton's Philosophy of Nature*. New York: Hafner

Thompson, P.: 1989, *The Structure of Biological Theories*. Albany, NY: SUNY Press

Toulmin, S.E.: 1963, *Foresight and Understanding*. New York: Harper Torchbooks

Toulmin, S.E.: 1972, *Human Understanding*, Vol. I. Princeton: Princeton University Press

Whewell, W.: 1949, *Of Induction*. London: Parker

Wimsatt, W.: 1974, 'Reductive Explanation: A functional Account', *Boston Studies in the Philosophy of Science: PSA 1974*, Cohen, R.S. (ed.) Dordrecht: D. Reidel Publishing Co.

Wimsatt, W.: 1976, 'Reductionism, Levels of Organization, and the Mind-Body Problem', in Globus, G. et. al. (eds.) *Brain and Consciousness: Scientific and Philosophical Strategies*. New York: Plenum Press

Wimsatt, W.: 1979, 'Reduction and Reductionism', in Asquith, P. and Kyburg, H. (eds.) *Current Research in Philosophy of Science*. East Lansing, MI: Philosophy of Science Association

Wimsatt, W.: 1980, 'Reductionistic Research Strategies and their Biases in the Units of Selection Controversy', in Nickles, T. (ed.) *Scientific Discovery: Case Studies*. Dordrecht: D. Reidel, 213–259

Wimsatt, W.: 1981, 'Robustness, Reliability and Multiple Determination in Science', in M. Brewer and B Collins, eds. *Knowing and Valuing in the Social Sciences: A Tribute to Donald T. Campbell*. San Francisco: Jossey-Bass

Wimsatt, W.: 1985, 'Heuristics and the Study of Human Behavior', in Fiske, D.W. and Shweder, R. (eds.) *Metatheory in Social Science*. Chicago: University of Chicago Press, 293–314

Winston, P.: 1977, *Artificial Intelligence*. Reading, MA: Addison-Wesley

Wittgenstein, L.: 1953, *Philosophical Investigations*. Oxford: Basil Blackwell

Wright, L.: 1989, *Practical Reasoning*. New York: Harcourt Brace Jovanovich

INDEX

129. M. W. Wartofsky, *Models*. Representation and the Scientific Understanding. [Boston Studies in the Philosophy of Science, Vol. XLVIII] 1979
ISBN 90-277-0736-7; Pb 90-277-0947-5
130. D. Ihde, *Technics and Praxis*. A Philosophy of Technology. [Boston Studies in the Philosophy of Science, Vol. XXIV] 1979 ISBN 90-277-0953-X; Pb 90-277-0954-8
131. J. J. Wiatr (ed.), *Polish Essays in the Methodology of the Social Sciences*. [Boston Studies in the Philosophy of Science, Vol. XXIX] 1979
ISBN 90-277-0723-5; Pb 90-277-0956-4
132. W. C. Salmon (ed.), *Hans Reichenbach: Logical Empiricist*. 1979
ISBN 90-277-0958-0
133. P. Bieri, R.-P. Horstmann and L. Krüger (eds.), *Transcendental Arguments in Science*. Essays in Epistemology. 1979 ISBN 90-277-0963-7; Pb 90-277-0964-5
134. M. Marković and G. Petrović (eds.), *Praxis*. Yugoslav Essays in the Philosophy and Methodology of the Social Sciences. [Boston Studies in the Philosophy of Science, Vol. XXXVI] 1979 ISBN 90-277-0727-8; Pb 90-277-0968-8
135. R. Wójcicki, *Topics in the Formal Methodology of Empirical Sciences*. Translated from Polish. 1979 ISBN 90-277-1004-X
136. G. Radnitzky and G. Andersson (eds.), *The Structure and Development of Science*. [Boston Studies in the Philosophy of Science, Vol. LIX] 1979
ISBN 90-277-0994-7; Pb 90-277-0995-5
137. J. C. Webb, *Mechanism, Mentalism and Metamathematics*. An Essay on Finitism. 1980 ISBN 90-277-1046-5
138. D. F. Gustafson and B. L. Tapscott (eds.), *Body, Mind and Method*. Essays in Honor of Virgil C. Aldrich. 1979 ISBN 90-277-1013-9
139. L. Nowak, *The Structure of Idealization*. Towards a Systematic Interpretation of the Marxian Idea of Science. 1980 ISBN 90-277-1014-7
140. C. Perelman, *The New Rhetoric and the Humanities*. Essays on Rhetoric and Its Applications. Translated from French and German. With an Introduction by H. Zyskind. 1979 ISBN 90-277-1018-X; Pb 90-277-1019-8
141. W. Rabinowicz, *Universalizability*. A Study in Morals and Metaphysics. 1979
ISBN 90-277-1020-2
142. C. Perelman, *Justice, Law and Argument*. Essays on Moral and Legal Reasoning. Translated from French and German. With an Introduction by H.J. Berman. 1980
ISBN 90-277-1089-9; Pb 90-277-1090-2
143. S. Kanger and S. Öhman (eds.), *Philosophy and Grammar*. Papers on the Occasion of the Quincentennial of Uppsala University. 1981 ISBN 90-277-1091-0
144. T. Pawlowski, *Concept Formation in the Humanities and the Social Sciences*. 1980
ISBN 90-277-1096-1
145. J. Hintikka, D. Gruender and E. Agazzi (eds.), *Theory Change, Ancient Axiomatics and Galileo's Methodology*. Proceedings of the 1978 Pisa Conference on the History and Philosophy of Science, Volume I. 1981 ISBN 90-277-1126-7
146. J. Hintikka, D. Gruender and E. Agazzi (eds.), *Probabilistic Thinking, Thermodynamics, and the Interaction of the History and Philosophy of Science*. Proceedings of the 1978 Pisa Conference on the History and Philosophy of Science, Volume II. 1981 ISBN 90-277-1127-5
147. U. Mönnich (ed.), *Aspects of Philosophical Logic*. Some Logical Forays into Central Notions of Linguistics and Philosophy. 1981 ISBN 90-277-1201-8
148. D. M. Gabbay, *Semantical Investigations in Heyting's Intuitionistic Logic*. 1981
ISBN 90-277-1202-6

106. K. Kosík, *Dialectics of the Concrete.* A Study on Problems of Man and World. [Boston Studies in the Philosophy of Science, Vol. LII] 1976
ISBN 90-277-0761-8; Pb 90-277-0764-2
107. N. Goodman, *The Structure of Appearance.* 3rd ed. with an Introduction by G. Hellman. [Boston Studies in the Philosophy of Science, Vol. LIII] 1977
ISBN 90-277-0773-1; Pb 90-277-0774-X
108. K. Ajdukiewicz, *The Scientific World-Perspective and Other Essays, 1931-1963.* Translated from Polish. Edited and with an Introduction by J. Giedymin. 1978
ISBN 90-277-0527-5
109. R. L. Causey, *Unity of Science.* 1977 ISBN 90-277-0779-0
110. R. E. Grandy, *Advanced Logic for Applications.* 1977 ISBN 90-277-0781-2
111. R. P. McArthur, *Tense Logic.* 1976 ISBN 90-277-0697-2
112. L. Lindahl, *Position and Change.* A Study in Law and Logic. Translated from Swedish by P. Needham. 1977 ISBN 90-277-0787-1
113. R. Tuomela, *Dispositions.* 1978 ISBN 90-277-0810-X
114. H. A. Simon, *Models of Discovery and Other Topics in the Methods of Science.* [Boston Studies in the Philosophy of Science, Vol. LIV] 1977
ISBN 90-277-0812-6; Pb 90-277-0858-4
115. R. D. Rosenkrantz, *Inference, Method and Decision.* Towards a Bayesian Philosophy of Science. 1977 ISBN 90-277-0817-7; Pb 90-277-0818-5
116. R. Tuomela, *Human Action and Its Explanation.* A Study on the Philosophical Foundations of Psychology. 1977 ISBN 90-277-0824-X
117. M. Lazerowitz, *The Language of Philosophy.* Freud and Wittgenstein. [Boston Studies in the Philosophy of Science, Vol. LV] 1977
ISBN 90-277-0826-6; Pb 90-277-0862-2
118. Not published
119. J. Pelc (ed.), *Semiotics in Poland, 1894–1969.* Translated from Polish. 1979
ISBN 90-277-0811-8
120. I. Pörn, *Action Theory and Social Science.* Some Formal Models. 1977
ISBN 90-277-0846-0
121. J. Margolis, *Persons and Mind.* The Prospects of Nonreductive Materialism. [Boston Studies in the Philosophy of Science, Vol. LVII] 1977
ISBN 90-277-0854-1; Pb 90-277-0863-0
122. J. Hintikka, I. Niiniluoto, and E. Saarinen (eds.), *Essays on Mathematical and Philosophical Logic.* 1979 ISBN 90-277-0879-7
123. T. A. F. Kuipers, *Studies in Inductive Probability and Rational Expectation.* 1978
ISBN 90-277-0882-7
124. E. Saarinen, R. Hilpinen, I. Niiniluoto and M. P. Hintikka (eds.), *Essays in Honour of Jaakko Hintikka on the Occasion of His 50th Birthday.* 1979
ISBN 90-277-0916-5
125. G. Radnitzky and G. Andersson (eds.), *Progress and Rationality in Science.* [Boston Studies in the Philosophy of Science, Vol. LVIII] 1978
ISBN 90-277-0921-1; Pb 90-277-0922-X
126. P. Mittelstaedt, *Quantum Logic.* 1978 ISBN 90-277-0925-4
127. K. A. Bowen, *Model Theory for Modal Logic.* Kripke Models for Modal Predicate Calculi. 1979 ISBN 90-277-0929-7
128. H. A. Bursen, *Dismantling the Memory Machine.* A Philosophical Investigation of Machine Theories of Memory. 1978 ISBN 90-277-0933-5

149. E. Agazzi (ed.), *Modern Logic – A Survey*. Historical, Philosophical, and Mathematical Aspects of Modern Logic and Its Applications. 1981 ISBN 90-277-1137-2
150. A. F. Parker-Rhodes, *The Theory of Indistinguishables*. A Search for Explanatory Principles below the Level of Physics. 1981 ISBN 90-277-1214-X
151. J. C. Pitt, *Pictures, Images, and Conceptual Change*. An Analysis of Wilfrid Sellars' Philosophy of Science. 1981 ISBN 90-277-1276-X; Pb 90-277-1277-8
152. R. Hilpinen (ed.), *New Studies in Deontic Logic*. Norms, Actions, and the Foundations of Ethics. 1981 ISBN 90-277-1278-6; Pb 90-277-1346-4
153. C. Dilworth, *Scientific Progress*. A Study Concerning the Nature of the Relation between Successive Scientific Theories. 2nd, rev. and augmented ed., 1986 ISBN 90-277-2215-3; Pb 90-277-2216-1
154. D. Woodruff Smith and R. McIntyre, *Husserl and Intentionality*. A Study of Mind, Meaning, and Language. 1982 ISBN 90-277-1392-8; Pb 90-277-1730-3
155. R. J. Nelson, *The Logic of Mind*. 2nd. ed., 1989 ISBN 90-277-2819-4; Pb 90-277-2822-4
156. J. F. A. K. van Benthem, *The Logic of Time*. A Model-Theoretic Investigation into the Varieties of Temporal Ontology, and Temporal Discourse. 1983; 2nd ed., 1991 ISBN 0-7923-1081-0
157. R. Swinburne (ed.), *Space, Time and Causality*. 1983 ISBN 90-277-1437-1
158. E. T. Jaynes, *Papers on Probability, Statistics and Statistical Physics*. Ed. by R. D. Rozenkrantz. 1983 ISBN 90-277-1448-7; Pb (1989) 0-7923-0213-3
159. T. Chapman, *Time: A Philosophical Analysis*. 1982 ISBN 90-277-1465-7
160. E. N. Zalta, *Abstract Objects*. An Introduction to Axiomatic Metaphysics. 1983 ISBN 90-277-1474-6
161. S. Harding and M. B. Hintikka (eds.), *Discovering Reality*. Feminist Perspectives on Epistemology, Metaphysics, Methodology, and Philosophy of Science. 1983 ISBN 90-277-1496-7; Pb 90-277-1538-6
162. M. A. Stewart (ed.), *Law, Morality and Rights*. 1983 ISBN 90-277-1519-X
163. D. Mayr and G. Süssmann (eds.), *Space, Time, and Mechanics*. Basic Structures of a Physical Theory. 1983 ISBN 90-277-1525-4
164. D. Gabbay and F. Guenthner (eds.), *Handbook of Philosophical Logic*. Vol. I: Elements of Classical Logic. 1983 ISBN 90-277-1542-4
165. D. Gabbay and F. Guenthner (eds.), *Handbook of Philosophical Logic*. Vol. II: Extensions of Classical Logic. 1984 ISBN 90-277-1604-8
166. D. Gabbay and F. Guenthner (eds.), *Handbook of Philosophical Logic*. Vol. III: Alternative to Classical Logic. 1986 ISBN 90-277-1605-6
167. D. Gabbay and F. Guenthner (eds.), *Handbook of Philosophical Logic*. Vol. IV: Topics in the Philosophy of Language. 1989 ISBN 90-277-1606-4
168. A. J. I. Jones, *Communication and Meaning*. An Essay in Applied Modal Logic. 1983 ISBN 90-277-1543-2
169. M. Fitting, *Proof Methods for Modal and Intuitionistic Logics*. 1983 ISBN 90-277-1573-4
170. J. Margolis, *Culture and Cultural Entities*. Toward a New Unity of Science. 1984 ISBN 90-277-1574-2
171. R. Tuomela, *A Theory of Social Action*. 1984 ISBN 90-277-1703-6
172. J. J. E. Gracia, E. Rabossi, E. Villanueva and M. Dascal (eds.), *Philosophical Analysis in Latin America*. 1984 ISBN 90-277-1749-4
173. P. Ziff, *Epistemic Analysis*. A Coherence Theory of Knowledge. 1984 ISBN 90-277-1751-7

174. P. Ziff, *Antiaesthetics*. An Appreciation of the Cow with the Subtile Nose. 1984 ISBN 90-277-1773-7
175. W. Balzer, D. A. Pearce, and H.-J. Schmidt (eds.), *Reduction in Science*. Structure, Examples, Philosophical Problems. 1984 ISBN 90-277-1811-3
176. A. Peczenik, L. Lindahl and B. van Roermund (eds.), *Theory of Legal Science*. Proceedings of the Conference on Legal Theory and Philosophy of Science (Lund, Sweden, December 1983). 1984 ISBN 90-277-1834-2
177. I. Niiniluoto, *Is Science Progressive?* 1984 ISBN 90-277-1835-0
178. B. K. Matilal and J. L. Shaw (eds.), *Analytical Philosophy in Comparative Perspective*. Exploratory Essays in Current Theories and Classical Indian Theories of Meaning and Reference. 1985 ISBN 90-277-1870-9
179. P. Kroes, *Time: Its Structure and Role in Physical Theories*. 1985 ISBN 90-277-1894-6
180. J. H. Fetzer, *Sociobiology and Epistemology*. 1985 ISBN 90-277-2005-3; Pb 90-277-2006-1
181. L. Haaparanta and J. Hintikka (eds.), *Frege Synthesized*. Essays on the Philosophical and Foundational Work of Gottlob Frege. 1986 ISBN 90-277-2126-2
182. M. Detlefsen, *Hilbert's Program*. An Essay on Mathematical Instrumentalism. 1986 ISBN 90-277-2151-3
183. J. L. Golden and J. J. Pilotta (eds.), *Practical Reasoning in Human Affairs*. Studies in Honor of Chaim Perelman. 1986 ISBN 90-277-2255-2
184. H. Zandvoort, *Models of Scientific Development and the Case of Nuclear Magnetic Resonance*. 1986 ISBN 90-277-2351-6
185. I. Niiniluoto, *Truthlikeness*. 1987 ISBN 90-277-2354-0
186. W. Balzer, C. U. Moulines and J. D. Sneed, *An Architectonic for Science*. The Structuralist Program. 1987 ISBN 90-277-2403-2
187. D. Pearce, *Roads to Commensurability*. 1987 ISBN 90-277-2414-8
188. L. M. Vaina (ed.), *Matters of Intelligence*. Conceptual Structures in Cognitive Neuroscience. 1987 ISBN 90-277-2460-1
189. H. Siegel, *Relativism Refuted*. A Critique of Contemporary Epistemological Relativism. 1987 ISBN 90-277-2469-5
190. W. Callebaut and R. Pinxten, *Evolutionary Epistemology*. A Multiparadigm Program, with a Complete Evolutionary Epistemology Bibliograph. 1987 ISBN 90-277-2582-9
191. J. Kmita, *Problems in Historical Epistemology*. 1988 ISBN 90-277-2199-8
192. J. H. Fetzer (ed.), *Probability and Causality*. Essays in Honor of Wesley C. Salmon, with an Annotated Bibliography. 1988 ISBN 90-277-2607-8; Pb 1-5560-8052-2
193. A. Donovan, L. Laudan and R. Laudan (eds.), *Scrutinizing Science*. Empirical Studies of Scientific Change. 1988 ISBN 90-277-2608-6
194. H.R. Otto and J.A. Tuedio (eds.), *Perspectives on Mind*. 1988 ISBN 90-277-2640-X
195. D. Batens and J.P. van Bendegem (eds.), *Theory and Experiment*. Recent Insights and New Perspectives on Their Relation. 1988 ISBN 90-277-2645-0
196. J. Österberg, *Self and Others*. A Study of Ethical Egoism. 1988 ISBN 90-277-2648-5
197. D.H. Helman (ed.), *Analogical Reasoning*. Perspectives of Artificial Intelligence, Cognitive Science, and Philosophy. 1988 ISBN 90-277-2711-2
198. J. Wolenski, *Logic and Philosophy in the Lvov-Warsaw School*. 1989 ISBN 90-277-2749-X

199. R. Wójcicki, *Theory of Logical Calculi.* Basic Theory of Consequence Operations. 1988 ISBN 90-277-2785-6
200. J. Hintikka and M.B. Hintikka, *The Logic of Epistemology and the Epistemology of Logic.* Selected Essays. 1989 ISBN 0-7923-0040-8; Pb 0-7923-0041-6
201. E. Agazzi (ed.), *Probability in the Sciences.* 1988 ISBN 90-277-2808-9
202. M. Meyer (ed.), *From Metaphysics to Rhetoric.* 1989 ISBN 90-277-2814-3
203. R.L. Tieszen, *Mathematical Intuition.* Phenomenology and Mathematical Knowledge. 1989 ISBN 0-7923-0131-5
204. A. Melnick, *Space, Time, and Thought in Kant.* 1989 ISBN 0-7923-0135-8
205. D.W. Smith, *The Circle of Acquaintance.* Perception, Consciousness, and Empathy. 1989 ISBN 0-7923-0252-4
206. M.H. Salmon (ed.), *The Philosophy of Logical Mechanism.* Essays in Honor of Arthur W. Burks. With his Responses, and with a Bibliography of Burk's Work. 1990 ISBN 0-7923-0325-3
207. M. Kusch, *Language as Calculus vs. Language as Universal Medium.* A Study in Husserl, Heidegger, and Gadamer. 1989 ISBN 0-7923-0333-4
208. T.C. Meyering, *Historical Roots of Cognitive Science.* The Rise of a Cognitive Theory of Perception from Antiquity to the Nineteenth Century. 1989 ISBN 0-7923-0349-0
209. P. Kosso, *Observability and Observation in Physical Science.* 1989 ISBN 0-7923-0389-X
210. J. Kmita, *Essays on the Theory of Scientific Cognition.* 1990 ISBN 0-7923-0441-1
211. W. Sieg (ed.), *Acting and Reflecting.* The Interdisciplinary Turn in Philosophy. 1990 ISBN 0-7923-0512-4
212. J. Karpiński, *Causality in Sociological Research.* 1990 ISBN 0-7923-0546-9
213. H.A. Lewis (ed.), *Peter Geach: Philosophical Encounters.* 1991 ISBN 0-7923-0823-9
214. M. Ter Hark, *Beyond the Inner and the Outer.* Wittgenstein's Philosophy of Psychology. 1990 ISBN 0-7923-0850-6
215. M. Gosselin, *Nominalism and Contemporary Nominalism.* Ontological and Epistemological Implications of the Work of W.V.O. Quine and of N. Goodman. 1990 ISBN 0-7923-0904-9
216. J.H. Fetzer, D. Shatz and G. Schlesinger (eds.), *Definitions and Definability.* Philosophical Perspectives. 1991 ISBN 0-7923-1046-2
217. E. Agazzi and A. Cordero (eds.), *Philosophy and the Origin and Evolution of the Universe.* 1991 ISBN 0-7923-1322-4
218. M. Kusch, *Foucault's Strata and Fields.* An Investigation into Archaeological and Genealogical Science Studies. 1991 ISBN 0-7923-1462-X
219. C.J. Posy, *Kant's Philosophy of Mathematics.* Modern Essays. 1992 ISBN 0-7923-1495-6
220. G. Van de Vijver, *New Perspectives on Cybernetics.* Self-Organization, Autonomy and Connectionism. 1992 ISBN 0-7923-1519-7
221. J.C. Nyíri, *Tradition and Individuality.* Essays. 1992 ISBN 0-7923-1566-9
222. R. Howell, *Kant's Transcendental Deduction.* An Analysis of Main Themes in His Critical Philosophy. 1992 ISBN 0-7923-1571-5

223. A. García de la Sienra, *The Logical Foundations of the Marxian Theory of Value.* 1992 ISBN 0-7923-1778-5
224. D.S. Shwayder, *Statement and Referent.* An Inquiry into the Foundations of our Conceptual Order. 1992 ISBN 0-7923-1803-X
225. M. Rosen, *Problems of the Hegelian Dialectic.* Dialectic Reconstructed as a Logic of Human Reality. 1993 ISBN 0-7923-2047-6
226. P. Suppes, *Models and Methods in the Philosophy of Science: Selected Essays.* 1993 ISBN 0-7923-2211-8
227. R. M. Dancy (ed.), *Kant and Critique: New Essays in Honor of W. H. Werkmeister.* 1993 ISBN 0-7923-2244-4
228. J. Woleński (ed.), *Philosophical Logic in Poland.* 1993 ISBN 0-7923-2293-2
229. M. De Rijke (ed.), *Diamonds and Defaults.* Studies in Pure and Applied Intensional Logic. 1993 ISBN 0-7923-2342-4
230. B.K. Matilal and A. Chakrabarti (eds.), *Knowing from Words.* 1993 ISBN 0-7923-2345-9
231. S.A. Kleiner, *The Logic of Discovery.* A Theory of the Rationality of Scientific Research. 1993 ISBN 0-7923-2371-8

Previous volumes are still available.

KLUWER ACADEMIC PUBLISHERS – DORDRECHT / BOSTON / LONDON

MIX
Papier aus verantwortungsvollen Quellen
Paper from responsible sources
FSC® C105338

If you have any concerns about our products,
you can contact us on
ProductSafety@springernature.com

In case Publisher is established outside the EU,
the EU authorized representative is:
Springer Nature Customer Service Center GmbH
Europaplatz 3, 69115 Heidelberg, Germany

Printed by Libri Plureos GmbH
in Hamburg, Germany